IMAGES
of America

Coal Mining in Jefferson County

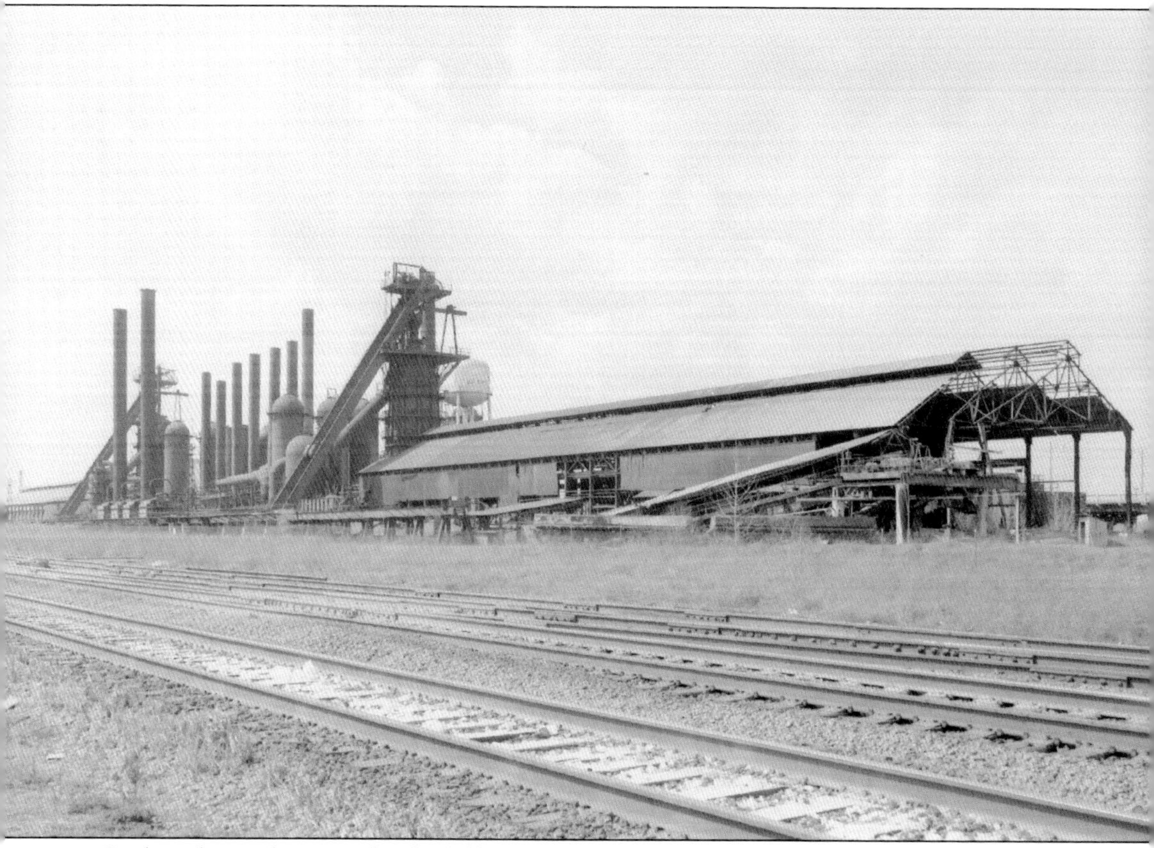

Coal machinery began to dot the Jefferson County landscape, and new mining towns like Praco, Labuco, Brookside, and Republic grew up. Not only did men and boys mine coal, but coal was loaded, coked, and then sent to Birmingham to fuel hungry blast furnaces for Sloss-Sheffield and Tennessee Coal Iron and Railroad. (Library of Congress.)

On the Cover: This photograph taken in 1910 in front of a mine at Bessie represents two facets of coal mining in Jefferson County. The men in the plain, dark clothing are free black miners. The others, dressed in white clothing, are just two of the many hundreds of convict coal miners who were leased by the state and local governments to work in area coal mines. In 1911, one year after this picture, over 120 convict miners were killed in an explosion at Bessie Mines. Following the explosion at Bessie, reformers sought to bring an end to the horrors of the convict lease system. (Birmingham Public Library [BPL] Archives Collection.)

IMAGES
of America

Coal Mining in Jefferson County

Staci Simon Glover

ARCADIA
PUBLISHING

ISBN 978-0-7385-8217-7

Published by Arcadia Publishing
Charleston, South Carolina

Printed in the United States of America

Library of Congress Control Number: 2010938145

For all general information, please contact Arcadia Publishing:
Telephone 843-853-2070
Fax 843-853-0044
E-mail sales@arcadiapublishing.com
For customer service and orders:
Toll-Free 1-888-313-2665

Visit us on the Internet at www.arcadiapublishing.com

This volume is dedicated to all those men and women who lived, worked, and built the towns that built Birmingham.

Contents

Acknowledgments

The idea for this volume emerged from a set of history talks presented in 2009 in Brookside, Alabama. The series focused on coal mining history and the history of Brookside. Participants were encouraged to bring family artifacts and photographs to share with others. Many responded, and the realization that no single pictorial volume on Jefferson County Coal Mining existed prompted the compilation of this volume. Thanks especially to the staff at Birmingham Public Library Archives Department and Debbie Pendelton with the Alabama Department of Archives and History. A special thanks to Kaye Nail, whose discussions of coal mining were an encouragement, and to Horace Defore, who shared with me photographs collected over a lifetime and gave me a starting point. Thanks to my mother, Carolyn Goode Simon, who gave constant support and encouragement. Thanks to my husband, Brian, and daughter Anna, who allowed me the time to spend working on this volume.

Most photographs in this volume come from various collections of the Library of Congress, Prints and Photographs Division. They will be noted as follows:

Library of Congress, Prints and Photographs Division, Farm Security Administration (LOC, FSA.)

Library of Congress, Prints and Photographs Division, National Child Labor Committee Collection (LOC, NCLC.)

Library of Congress, Prints and Photographs Division, Detroit Publishing Collection (LOC, DPC.)

Library of Congress, Prints and Photographs Division, Historic American Building Survey (LOC, HABS.)

Other photographs are from the Birmingham Public Library Archives (BPL Collection), the Brookside History Museum (BHM), the United States Geological Survey (USGS), and personal collections.

Introduction

Jefferson County, Alabama, was an agricultural outpost for Native Americans and yeomen farmers in the early 19th century. Created by the state legislature, the county that would become the industrial hub for the state by the end of the 19th century was created on December 13, 1819, just one day before Alabama became a state. Until 1821, Carrollsville served as the county seat; it was then moved to Elyton, where it remained until 1873 when it was moved to its present location in Birmingham. Traditionally, Elyton served the Native American population as a trading center, and as whites settled the area, many of the county's first settlers received land grants for their service in the War of 1812, many having fought with Andrew Jackson at Horseshoe Bend. The Tennessee Volunteers, led by Davey Crockett, traveled through the county, and impressed with the land they saw, many began to settle in the regions around Elyton, Pinson, Lipscomb, and Brookside.

Situated at the end of the Appalachian Mountain Range, Jefferson County sat in the middle of Alabama's hill country and had few plantations, as the soil types and hilly terrain were not favorable for planting cotton on a large scale. Those who settled the county primarily engaged in subsistence farming and a few owned slaves, though slave ownership was not practiced on a large scale. The county steadily grew an agricultural economy with crops such as wheat, corn, vegetables, and some cotton being marketed from small farms. Native Americans lived along the county's riverbanks and creek banks, carving symbols along the rock faces. They planted large gardens along Five Mile Creek and developed systems of irrigation to harness the creek's water to nourish their crops. Even today, when heavy rains wash over the creek's banks, Native American artifacts such as pottery and arrowheads may be found. In 1830, Pres. Andrew Jackson signed the Indian Removal Act, which began the process of removing Native Americans from southern lands and placing them on reservations in the West. In 1838, the Cherokee began their march west with some 4,000 dying along the way. The Trail of Tears, as the march became known, passed through Jefferson County and future coal mining country. With Native Americans displaced, the county resumed its agricultural bent toward small family farms occupied by native-born whites.

Though the Warrior Coal Field had been discovered in the early part of the century, coal veins remained relatively untapped until late in the century. During the Civil War, most of the South's coal came from the mines of Jefferson County and the outlying Birmingham Industrial District. However, after the war's conclusion, little emphasis was placed on coal mining as a key industry for immediate postwar economic development. After the founding of Birmingham, sitting along a transportation corridor that is still vital to the state's economy today, the discovery of iron ore, coal, and limestone in close proximity to one another offered entrepreneurs an obvious opportunity for industrial investment. Men like Daniel Pratt, James W. Sloss, Henry Debardelaben, and Enoch Ensley incorporated companies, sought investors, and equipped their firms with the leading technology of the time. However, one serious issue may have left them scratching their heads.

Industrialists seemed to have thought of everything from financing to new technology when setting up their businesses. However, they faced a predicament that they might not have expected:

how to attract a labor force. The South after the Civil War was in need of recapitalizing, building up and modernizing, and being brought into an industrial economy. Many whites in Jefferson County still toiled as yeoman farmers, scratching out a living on small farms either as owners or as tenants. Industrialists believed erroneously that these men would be willing to put aside their farms and work for a steady wage in the coal and iron ore mines or in one of Birmingham's iron foundries or steel mills. However, the agricultural clock—rising at dawn, breaking for lunch when one pleased, and stopping work when the sun went down—was hard for these farmers to give up. The industrial clock, governed by the piercing call of the whistle, was foreign to these agricultural workers. Industrial workers had little control over their days. They arrived at work when the whistle called, they stopped for lunch when the whistle blew again, and they left for home when the whistle sounded at the end of the day. Though many did eventually find jobs in the industrial work force, industrialists still had to find workers. They looked to Europe to fill that void.

Industrial boosters, traveling to cities and ports in the United States and across Europe, touted the riches that the Birmingham Industrial District had to offer willing workers. Recruiters praised Alabama's climate, schools, and the availability of jobs at fair wages, and representative democracy appealed to many European men seeking an escape from military conscription. Historians of immigration and migration patterns, such as John Bodnar, observe a push-pull or magnet feature to immigrant history. While the pull of America was multifold—promise of jobs, freedom from conscription, and the hope of making a better life—the push feature was equally as strong. Famine, conversion of farmland to industrial use, forced military service, and general unrest pushed millions of immigrants out of Europe and into America's industrial labor force.

Traveling to the United States—entering at New York, Charleston, or New Orleans—could be a daunting experience for immigrants. Many were surprised at the conditions they encountered upon arriving in the Birmingham Industrial District. One woman made the observation that she left cobblestone roads in Europe for the "muddy roads" of the new, bustling, and dangerous city that had only recently been carved out of the craggy Alabama hill country. Many native whites looked upon the new immigrants with disdain. Many immigrants from Russia and the Austro-Hungarian Empire dressed differently, women particularly, who dressed in long skirts and head coverings. United States naturalization records describe many of these male immigrants as dark or swarmy complexioned. In the racist Jim Crow South, many whites equated dark complexions with inferiority and danger. Moreover, many pooled all southern and eastern Europeans into one large group, failing to learn their unique differences. One young Brookside girl of Slovak descent experienced racism when a classmate teased her for her dark complexion and Slovak language.

Many immigrants to the Alabama coalfields were experienced with machinery used in Jefferson County mines. Industrial archaeologist Jack Bergstresser wrote that the mines owned by Sloss-Sheffield Iron and Steel Company were among the most mechanized in the district. It is not surprising then that Brookside and her surrounding towns, Cardiff and Blossburg, were dominated by immigrants from Slovakia, France, Scotland, and other foreign environs. In Brookside, Slovaks made up 37 percent of the households in the small town. Italians comprised the largest segment of Birmingham's population by 1910. While most worked in the large, company-owned mines, others worked small businesses and family-owned wagon mines.

Another source of labor for the mines in Jefferson County was the local prisons. Late in the 19th century, the state of Alabama and the local coal operators developed a system in which coal operators paid the state a fee for convict labor. Each side deemed the arrangement beneficial for both sides. Coal companies paid a much smaller fee to rent labor than was paid to regular employees, and the state collected funds that more than offset the cost of incarceration. Those who violated vagrancy laws and could not pay fines or court costs or committed other crimes would be candidates for incarceration and then subject to being placed in the convict lease system. While most of those caught up in the convict lease system were black, there was a segment of white males as well. One white man could not pay his court fees and was imprisoned for a brief sentence. However, the sentence of a few months became a death sentence as he was forced to work in the mines as a convict and never returned home to his family. In 1911, when Banner Mines blew, 128 miners lost their lives, 125 of them black convicts. The system was outlawed in the first decades of the 20th century.

A final source of labor for the coal operators was children. Young boys served as greasers and worked the coal cars as they approached the tipple for emptying, both dangerous and low-paying jobs. One greaser became famous when Lewis Wickes Hines snapped his photograph at Bessie Mines in the northern part of the county. Henry Sharpe Higginbotham, known to the nation as "Shorpy," claimed he was 12 at the time his photograph was taken in 1910. Hines, a photographer for the National Child Labor Committee, doubted the boy's claim and believed him to be several years younger. One of Hines's photographs shows Shorpy covered in grease carrying a heavy pail of oil. His job was to make sure all the moving parts of the heavy coal cars were greased and operated smoothly, a dangerous vocation since the heavy cars could roll over a child at any time. Shorpy's photograph was published widely and helped to lead to child labor laws to protect children from long working hours and dangerous jobs. However, many children aided their families in privately owned wagon mines. Wilbur Goode remembered digging out the faces of mines with a pick and shovel at the age of 13.

The heyday of the Jefferson County coal miner ended around 1920 when miners went on strike over wage reductions. Coal operators took the opportunity to close old mines and reduce the workforce. After 1920, many coal towns began to feel an economic depression that would not affect the rest of the country until the stock market crash at the end of the decade. Towns like Brookside, Cardiff, Labuco, Pinckney City, and others experienced an outward migration that forever changed their social and economic landscapes. Many miners moved to northern environs like Illinois. Several members of the Black family of Blossburg moved to Illinois to find work, never returning to their birthplace. Others who stayed faced hard times in which some resorted to making distilled liquor during Prohibition to make ends meet. They used the abandoned mines to hide their stills and store their liquor. One woman in Cardiff had a false wall built in her kitchen so that she could store her home brew behind it and deceive the tax revenue agents who regularly paid her a visit. Others placed their stills behind the pools of water that routinely accumulated in the mines so they could hear the sloshing footsteps of agents as they hunted the lawbreakers. While some made illegal whiskey to survive mine closures, many others went to work at the steel mills in Birmingham and Fairfield. Others in the Brookside area found work at an explosives plant built at Watson by the DuPont de Nemours Company.

Despite labor trouble, strikes, and mine closures, Jefferson County's coal miners left a legacy of hard work and pride. Without Birmingham's outlying coal communities, the city might never have become the "Magic City." It was upon the backs of the county's coal miners that the iron and steel industry was built. Through these pages, the photographs selected depict the mines in which the men and boys worked, the workers themselves, the communities in which they lived, and how they lived with their families.

One

Mines

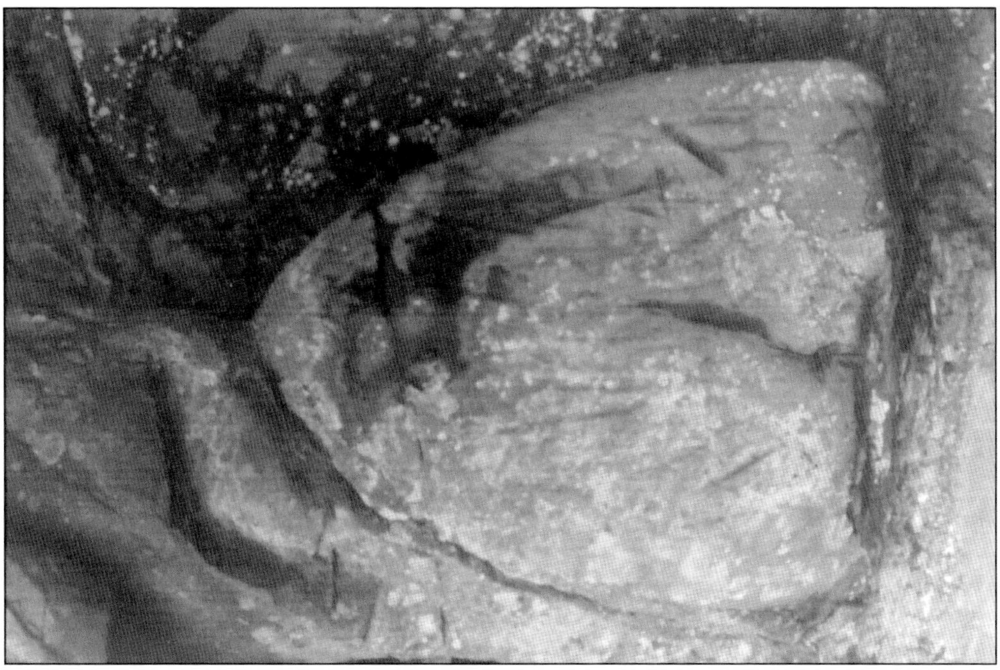

Long before coal mines dotted the countryside of Jefferson County, Native American camps dominated the area. After the Trail of Tears, all that was left of them were markings like these on rocks and in caves across the area. (Brookside History Museum, Albert Billits Collection.)

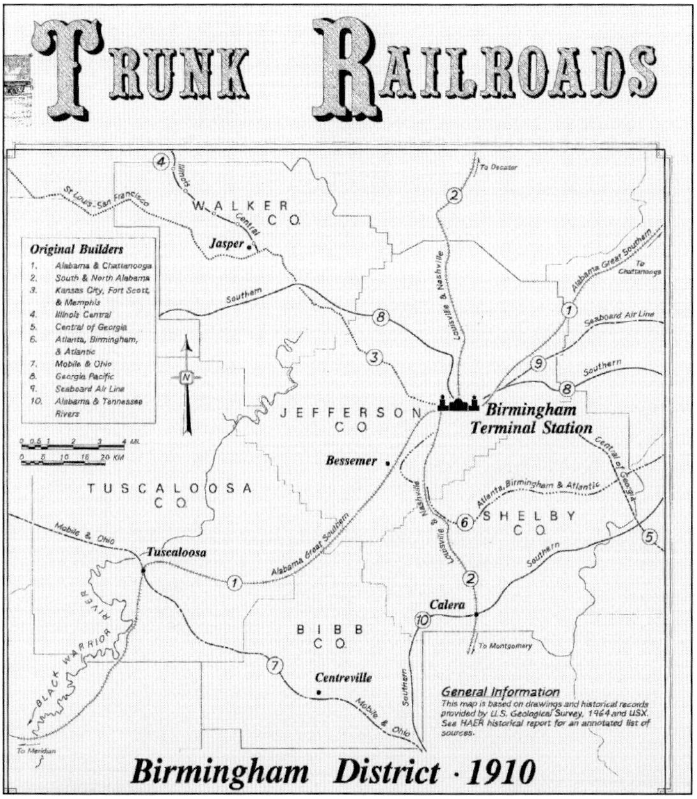

Railroad lines crisscrossed the Birmingham Industrial District, creating an adequate transportation hub for the growing iron, steel, and coal industries. Moreover, the L&N rail line bisected the state, and many of the coal towns of northern Jefferson County were serviced by it. Miners used these rail lines to travel from rural coal towns to Birmingham for supplies. (LOC, HABS.)

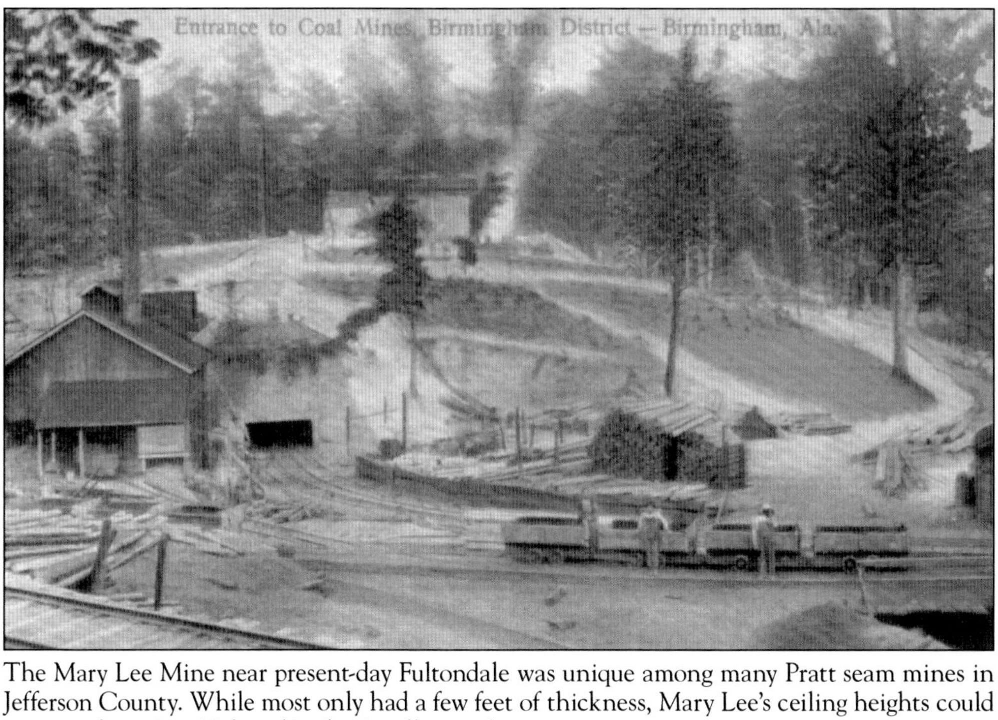

The Mary Lee Mine near present-day Fultondale was unique among many Pratt seam mines in Jefferson County. While most only had a few feet of thickness, Mary Lee's ceiling heights could measure from 4 to 10 feet. (Author's collection.)

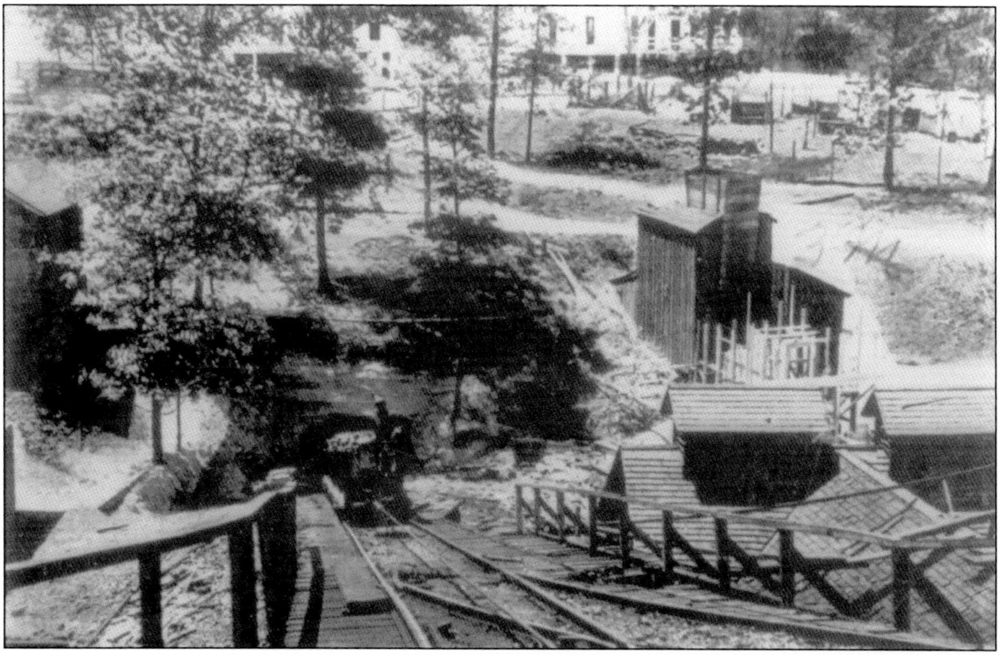

Coal mining began in the state prior to the Civil War, yet only when industrialists became interested in iron and steel production was coal truly considered a viable industry. Out of bucolic country landscapes, holes were blasted into the sides of mountains, machinery erected, and coal mining commenced. (Author's collection.)

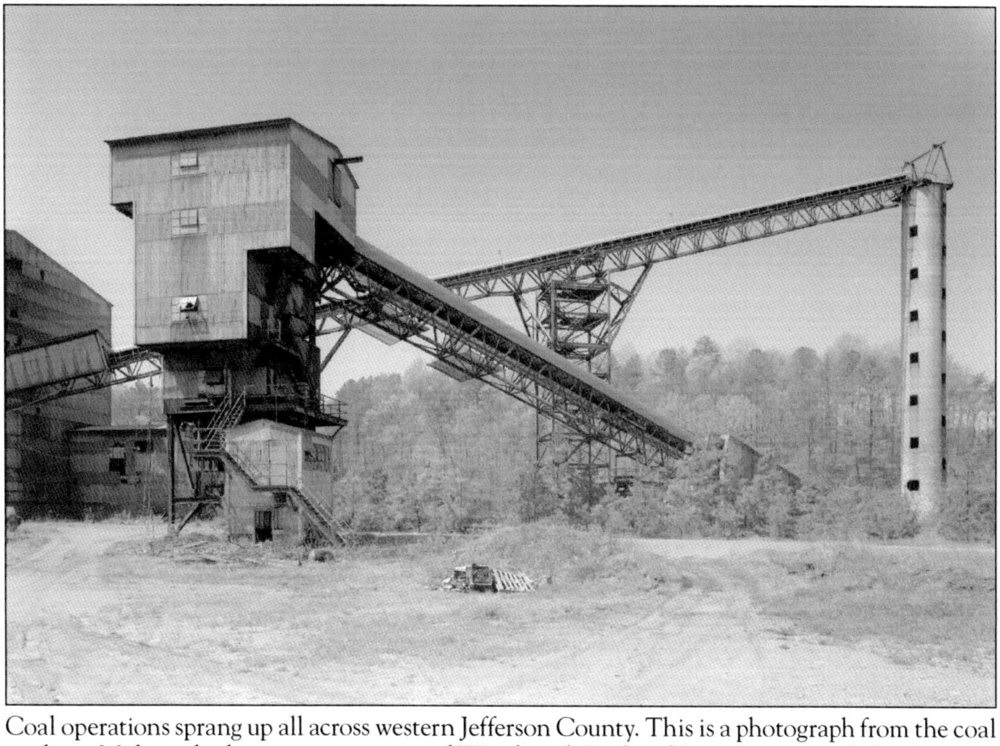

Coal operations sprang up all across western Jefferson County. This is a photograph from the coal works at Mulga, which was an operation of Woodward Coal and Iron Company. (LOC.)

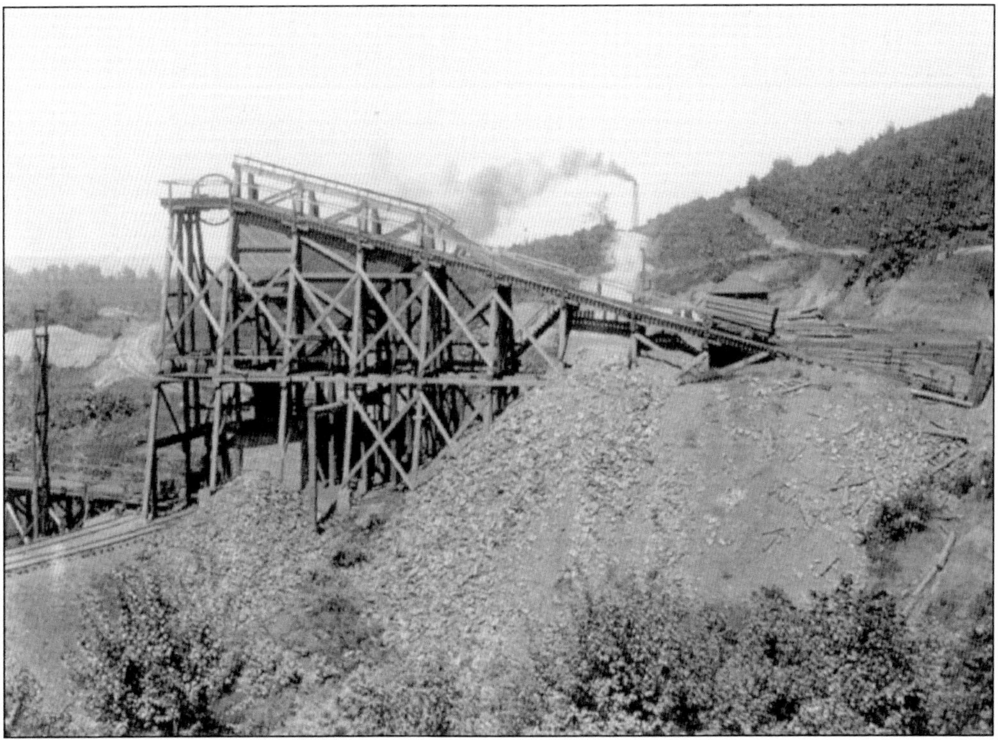

The Alice Mine was one of the first to open in the Birmingham Industrial District. (USGS.)

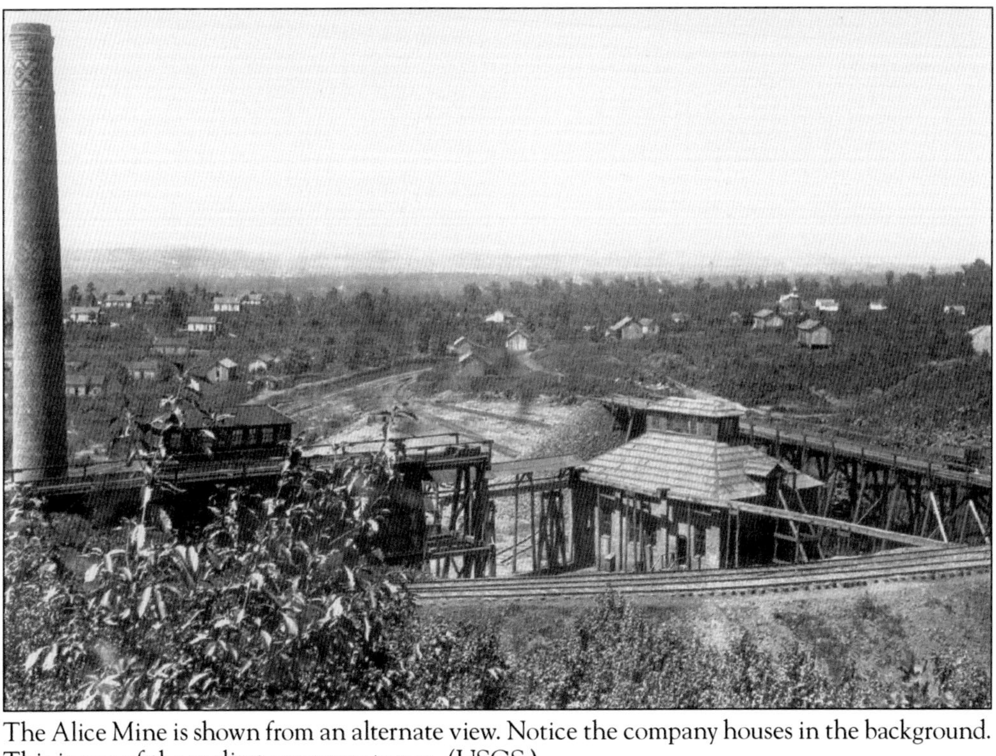

The Alice Mine is shown from an alternate view. Notice the company houses in the background. This is one of the earliest company towns. (USGS.)

Wagon mines, such as this one near Cardiff, were prevalent throughout Jefferson County. These mines were the smallest of operations, generally owned by families. Wagon mine operators mined coal for their personal use and supplied coal for households in and around Birmingham. Some provided coal for houses as far away as Highland Avenue. (Rosemary Nail.)

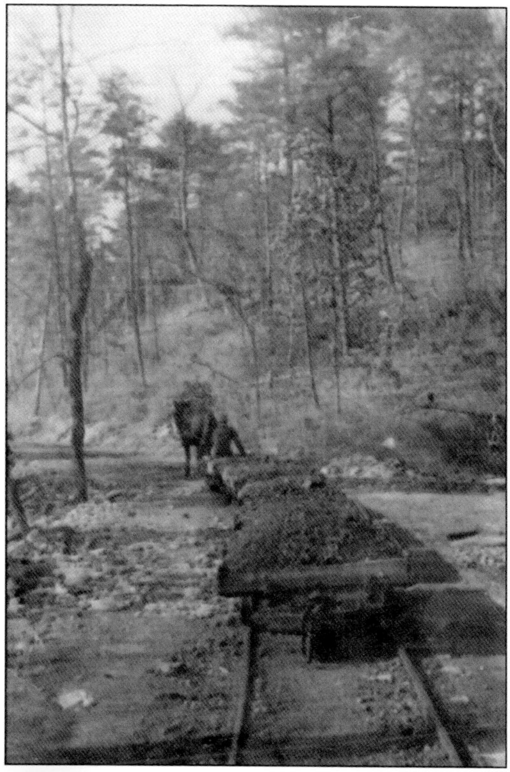

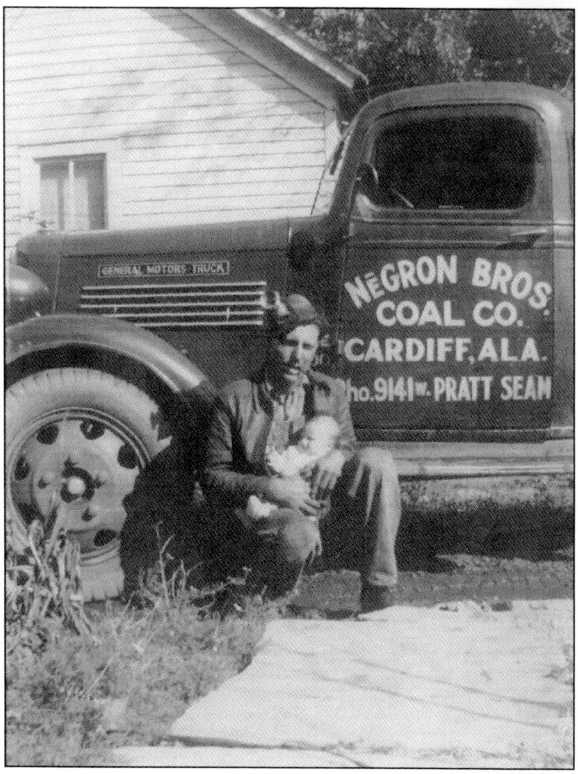

Negron Brothers Coal Company delivered coal from the family mines in Cardiff. The truck proudly displays Pratt Seam Coal. Early geologists deemed Pratt Seam Coal to be among the best in the state for coking purposes. (BHM.)

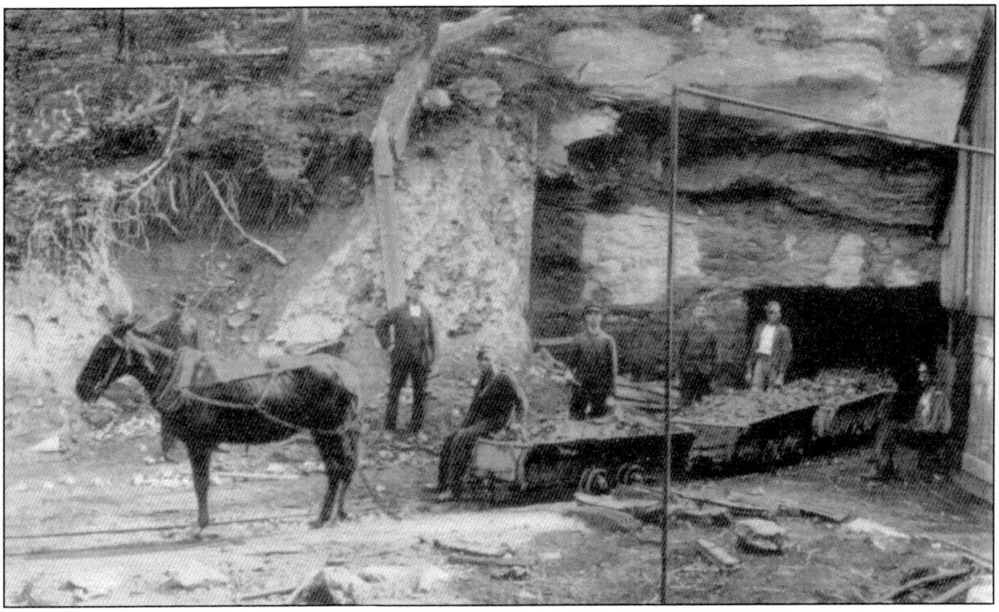

Though coal operators tried to use race to divide miners, race played no part when in the depths of a coal mine. This 1908 postcard of a wagon mine shows three white and three black coal miners as they prepare to off-load their cart. One miner said, "Everyone was the same in the mines." (Author's collection.)

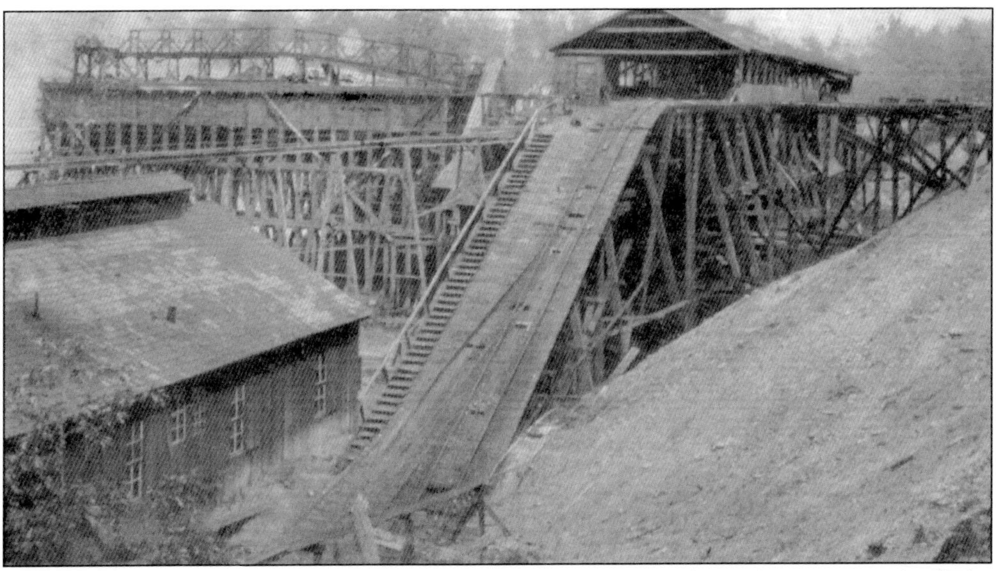

As corporate coal operations grew, they developed their rural industrial outposts with all the machinery necessary to remove coal from the ground, wash it, fire it, and turn it into coke. Part of the infrastructure, tipples earned their name easily as the structure from which coal cars were tipped and coal dumped into railroad cars. (Author's collection.)

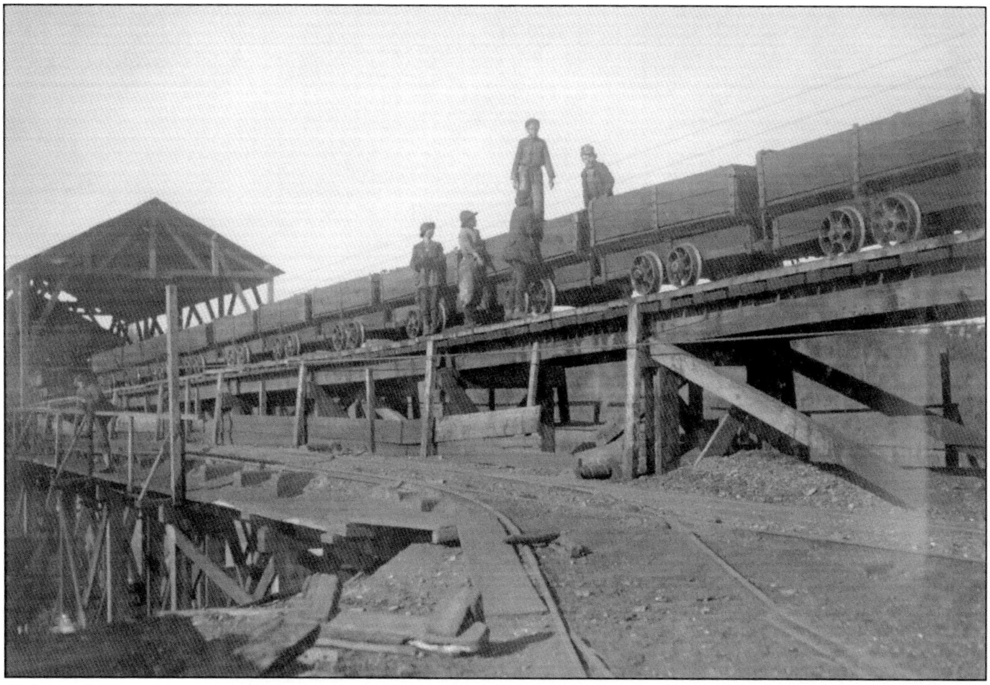

Often times, young boys worked the coal cars as they made their way up the tipple. Attending coal cars on the tipple required workers to grease car wheels and keep cars properly aligned. This was dangerous work since the cars were always moving and loaded with coal. (LOC, NCLC.)

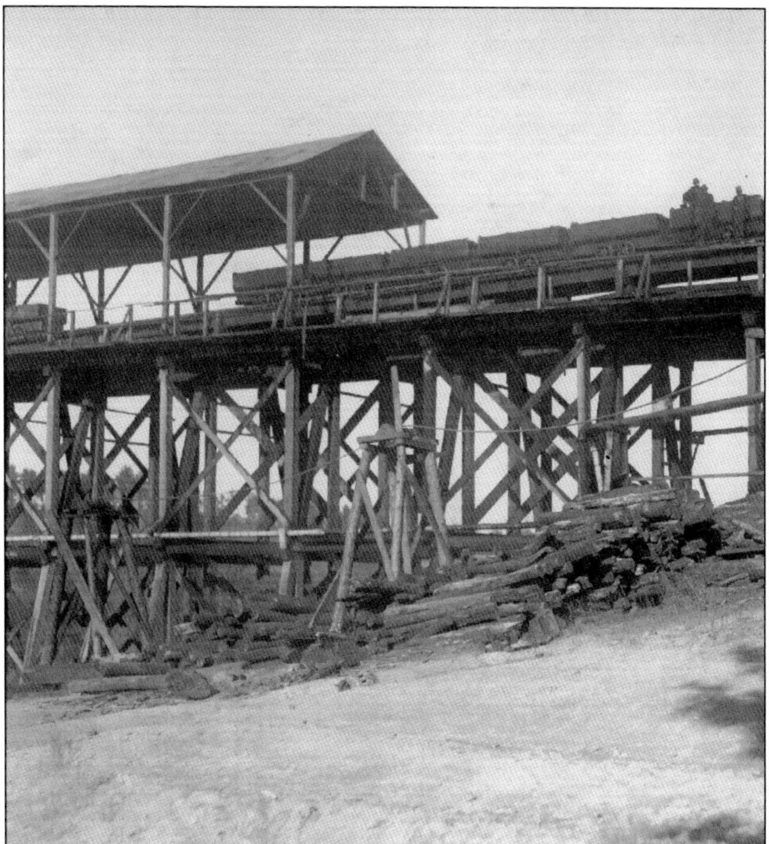

Other than inside the mine, the tipple was considered to be the most dangerous place to work due to all the moving cars. (LOC, NCLC.)

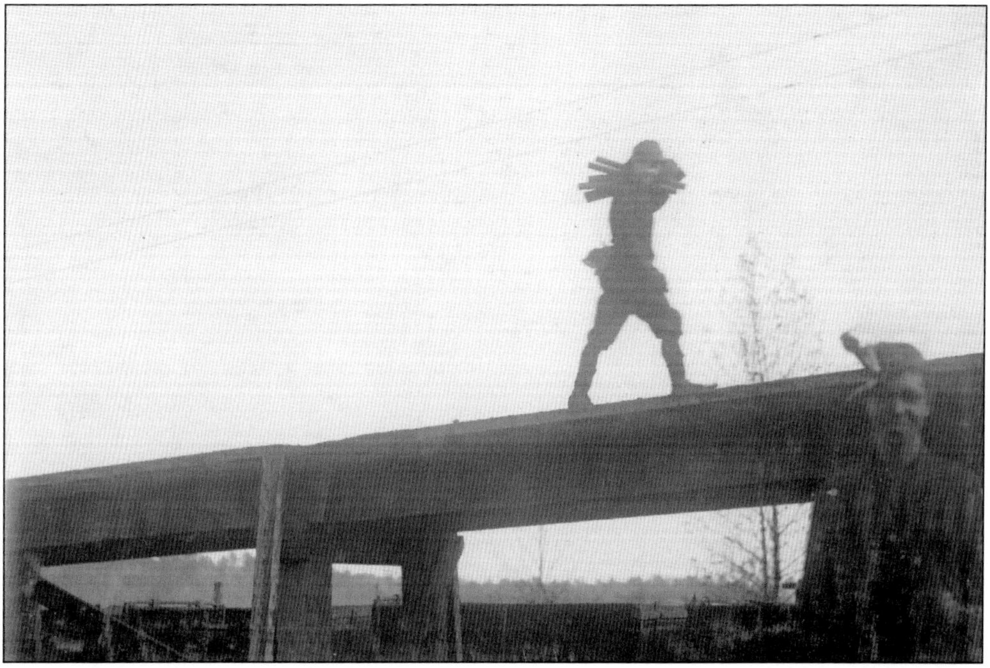
This young miner carries steel rods across the tipple at Bessie. (LOC, NCLC.)

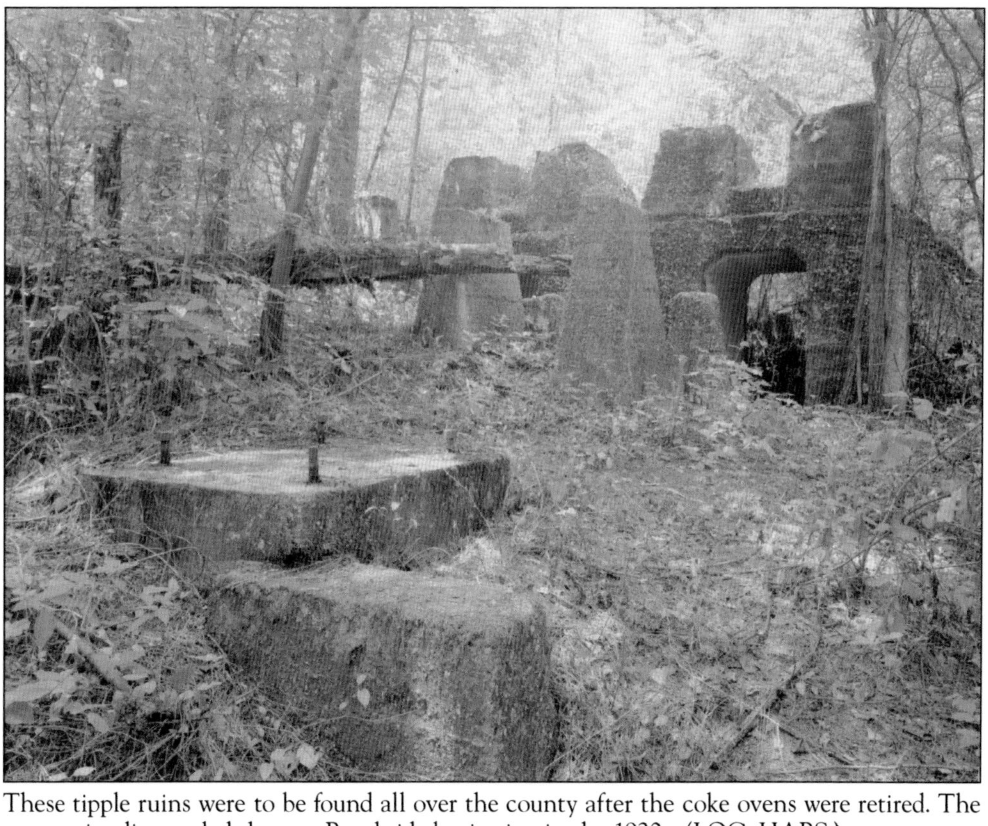
These tipple ruins were to be found all over the county after the coke ovens were retired. The companies dismantled them at Brookside beginning in the 1920s. (LOC, HABS.)

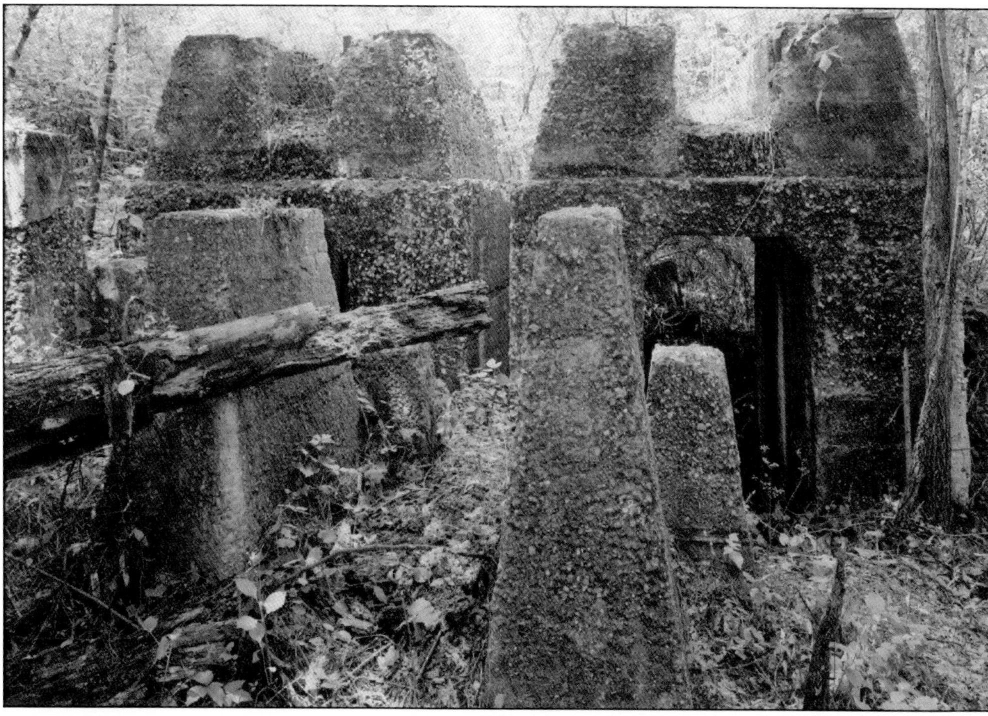

Some of the superstructure of Brookside's tipples, though dismantled in the 1920s, leaving only the foundation ruins, were used to build the entryway to the town school constructed in 1923. That structure stands today marking entrance to a local cemetery. (LOC, HABS.)

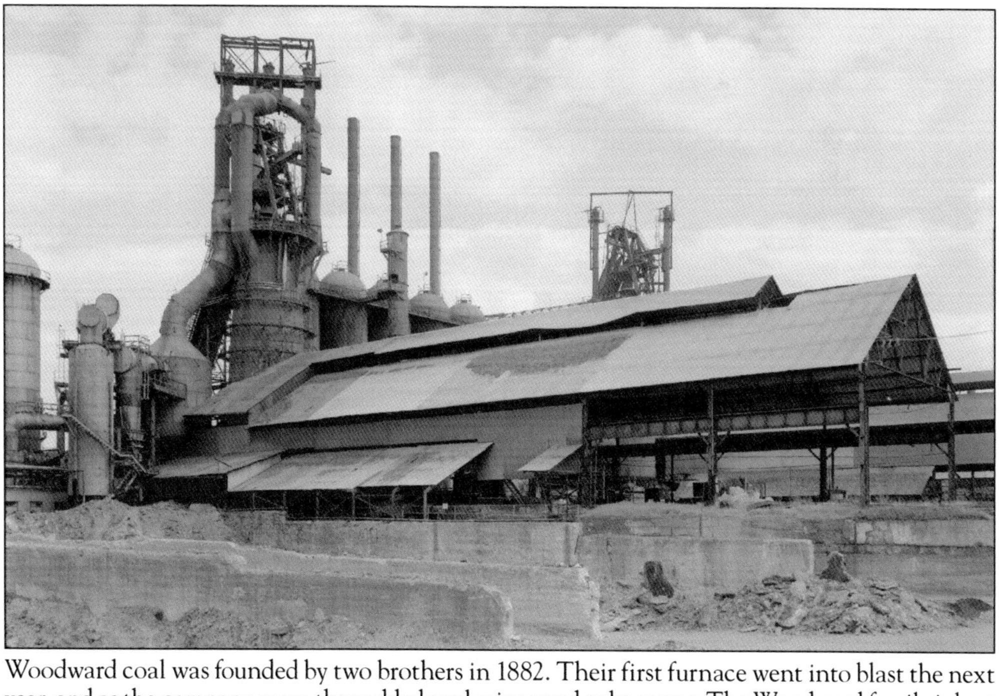

Woodward coal was founded by two brothers in 1882. Their first furnace went into blast the next year, and as the company grew, they added coal mines and coke ovens. The Woodward family is best known for their affiliation with building Rickwood Field, America's oldest ballpark. (LOC.)

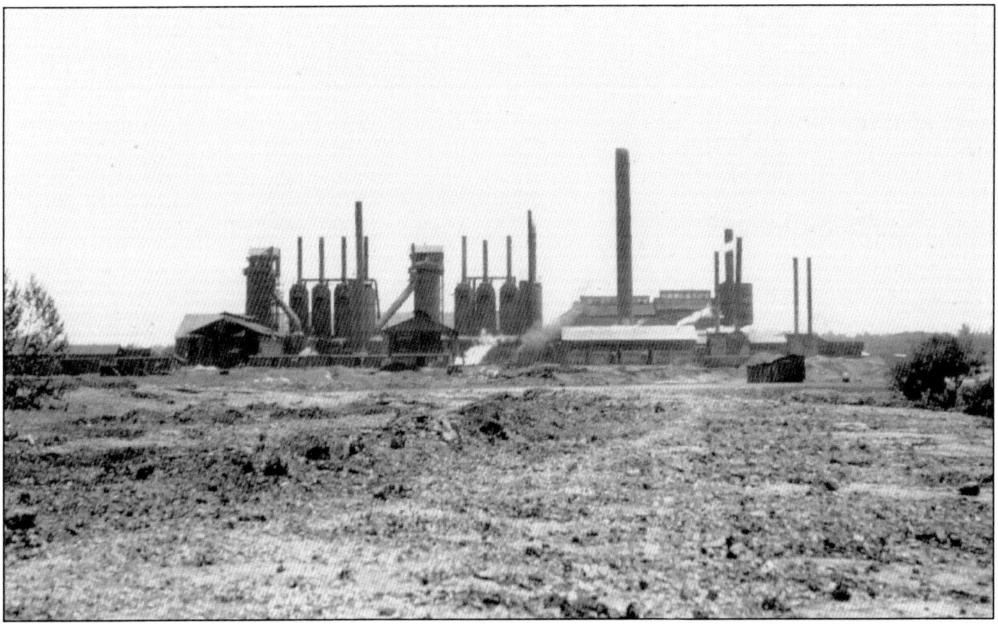

Here, the Sloss-Sheffield blast furnace is depicted. The Sloss Company owned many of the mines in the northern portion of the county to feed their furnaces so that pig iron could be produced. (USGS.)

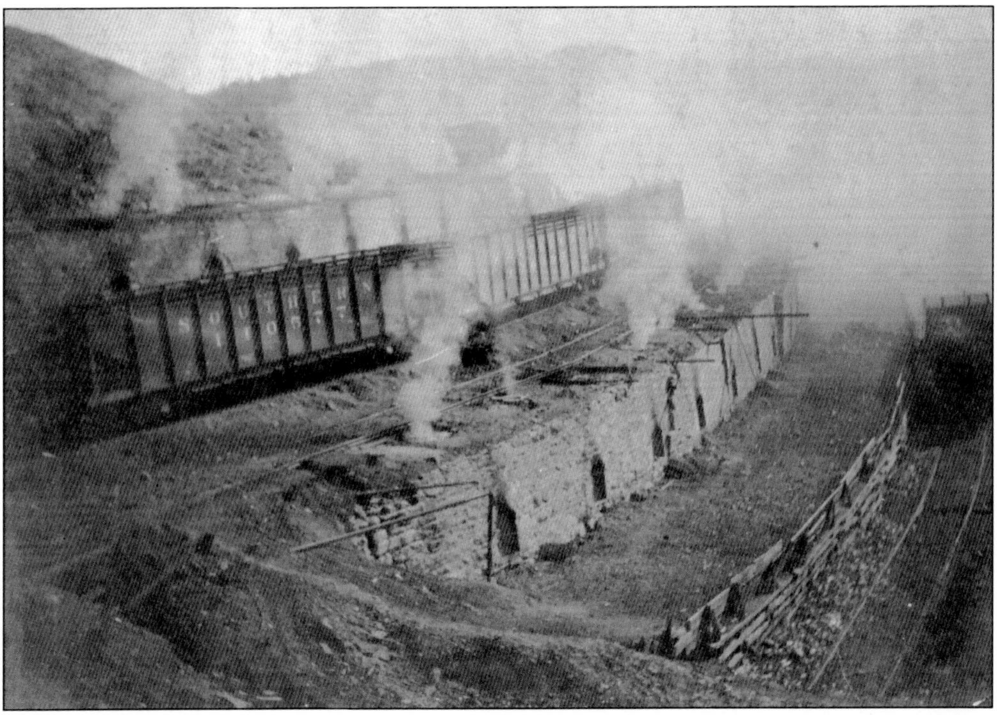

Beehive coke ovens were integral to the processing of coal for industrial application. Called beehive ovens because they resembled the arc of a beehive, they were constructed of fire brick that could withstand temperatures of over 2,000 degrees. The ovens roared both day and night, adding a glow to the night sky. (Author's collection.)

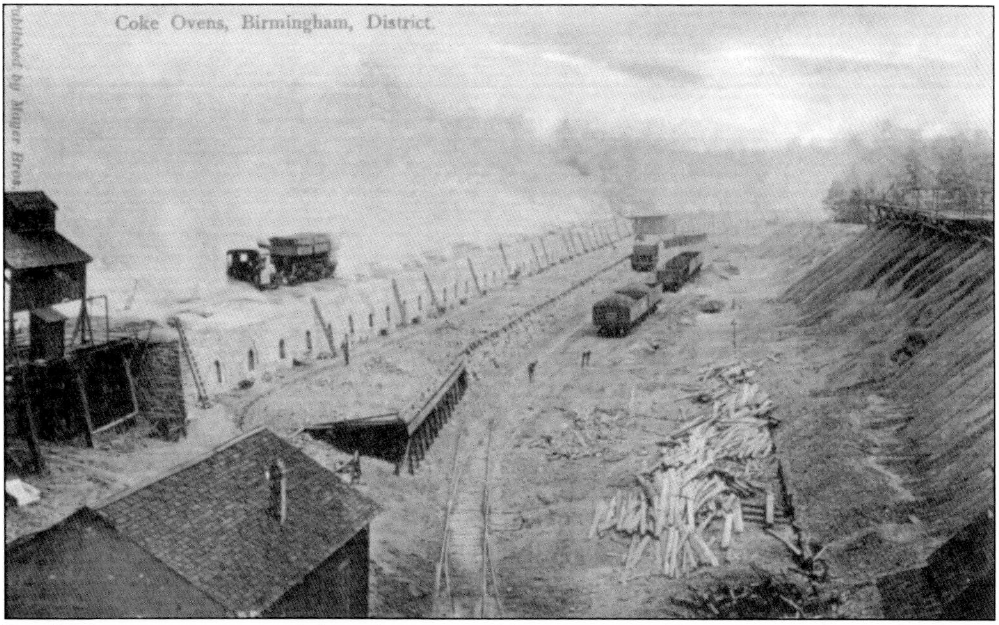

Brookside, sitting atop one of the richest parts of the Warrior Field, was home to some of the most mechanized mines in the district. Batteries of coke ovens wrapped around the hillside. The smell of sulfur and the release of chemicals filled the air and killed the vegetation in the surrounding area. Sulfur is now known to be one of leading causes of acid rain. (Author's collection.)

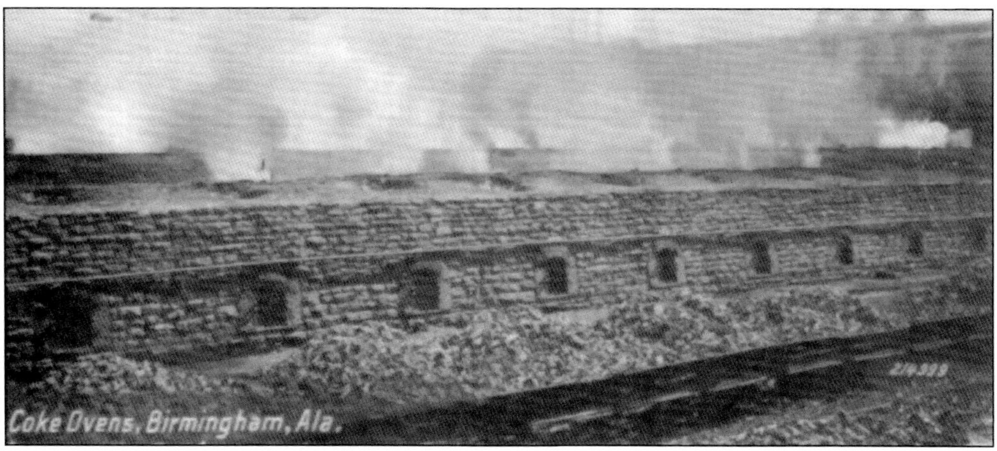

Coke ovens were built individually and then into batteries that might contain hundreds of ovens. Noxious fumes and chemicals pealed out of the ovens, killing vegetation in the nearby area. Only after the coke oven batteries were no longer used did vegetation once again begin to grow in the county's coal towns. (Author's collection.)

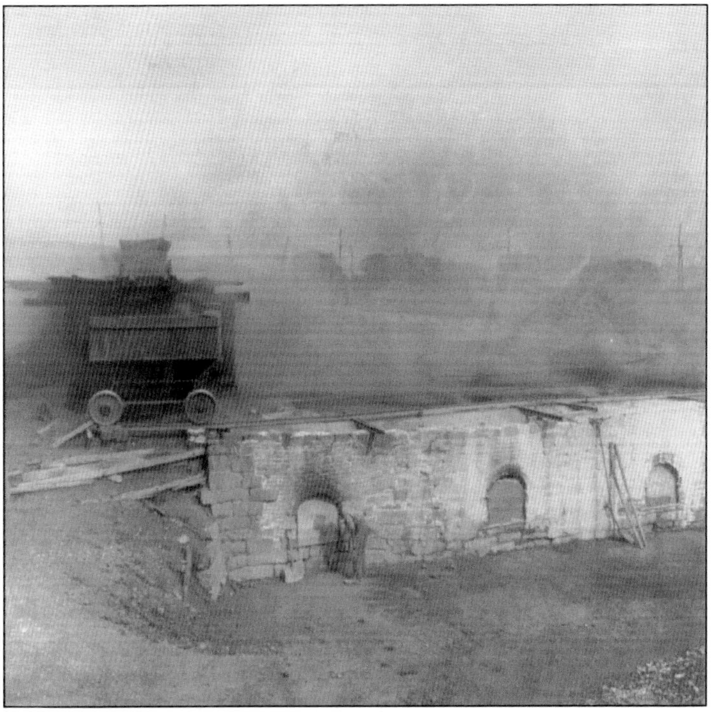

While many of the Lewisburg coke ovens had been abandoned in the 1920s, the need for increased production in 1941 forced Sloss-Sheffield to recommission 87 coke ovens in Lewisburg. While the battery included over 400 ovens, time had taken its toll and rendered 87 inoperable. (LOC.)

Coke ovens fell out of use by the 1920s and 1930s, as most coal was sent to coking facilities like the ABC coke plants in Tarrant and East Thomas. (LOC, HABS.)

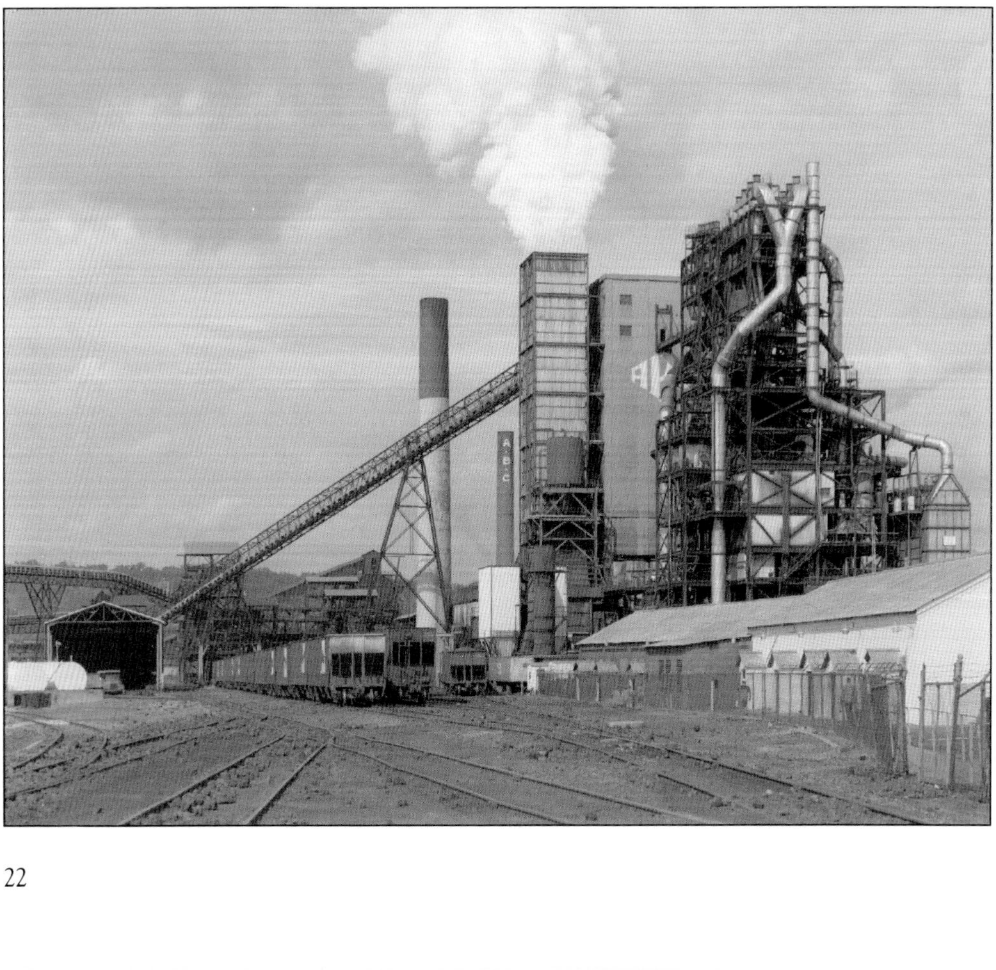

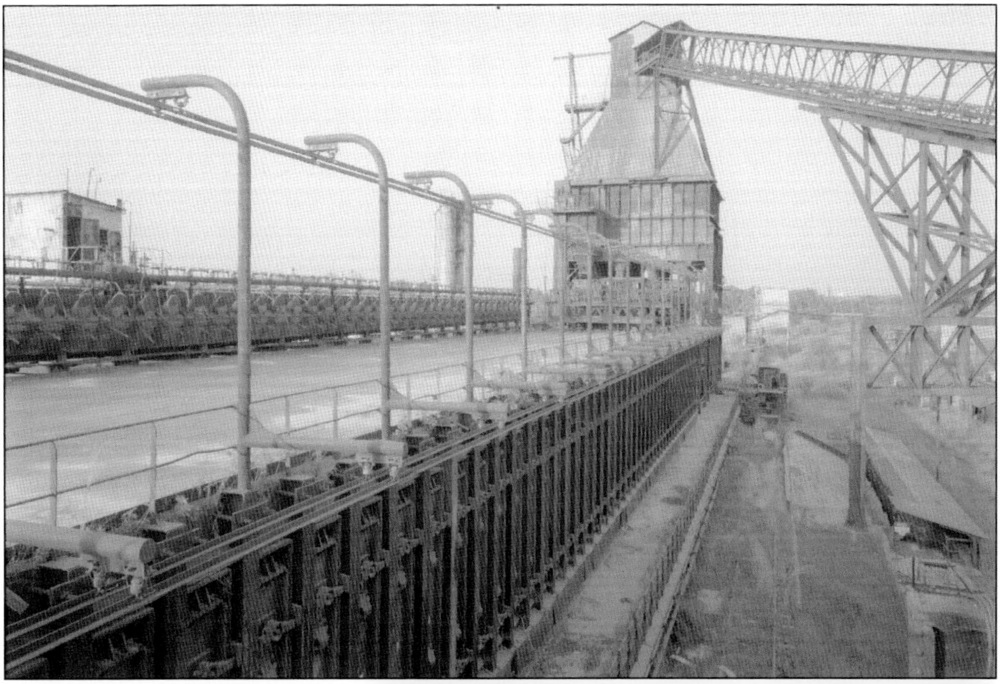

The coke works at Thomas proved to be one of the most technologically advanced sites in the district. This is a view of the coke ovens. They strike quite a sharp contrast to the on-site coke ovens at coal mining operations. (LOC, HABS.)

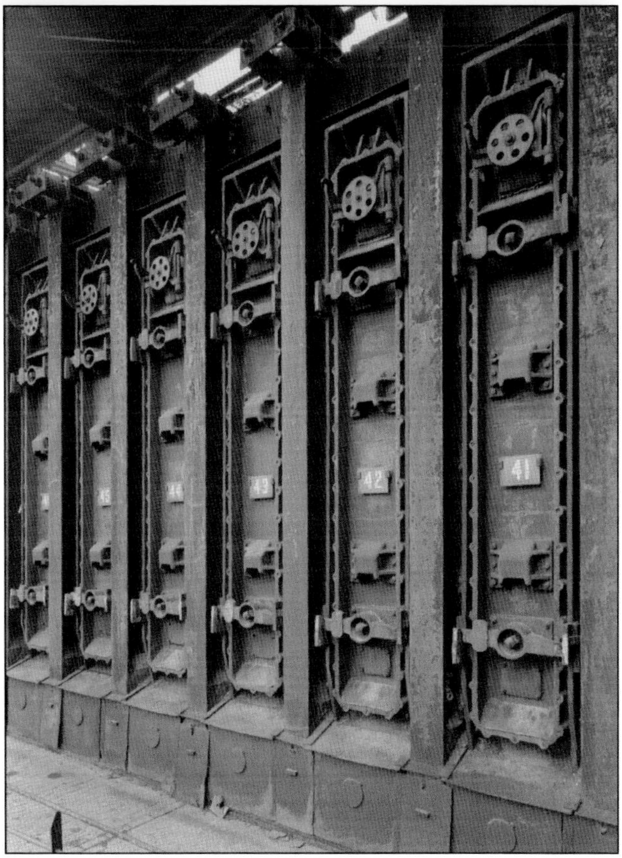

This is another view of the Thomas coke works. Rather than brick faces and doorways, these are metal. (LOC, HABS.)

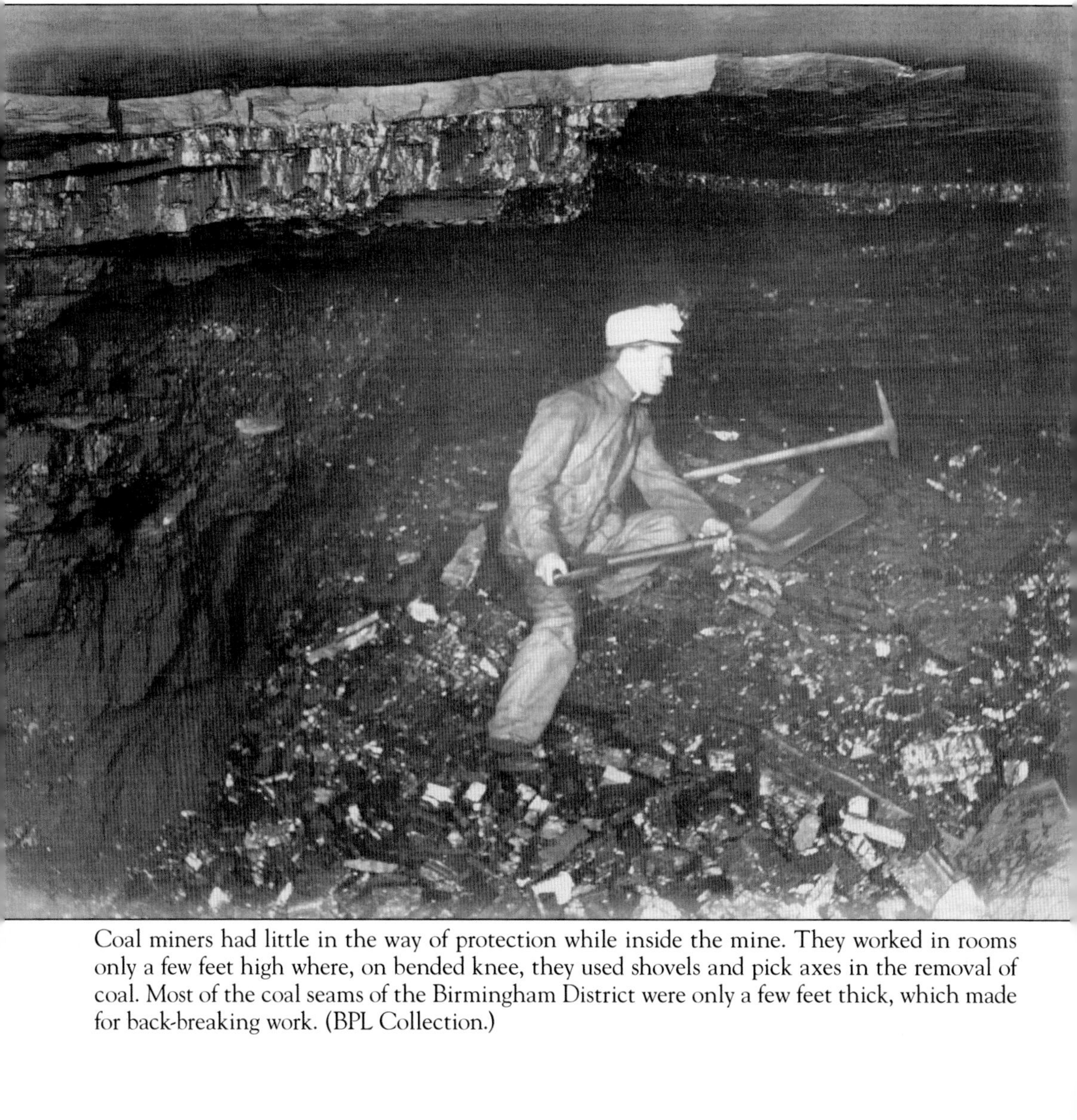

Coal miners had little in the way of protection while inside the mine. They worked in rooms only a few feet high where, on bended knee, they used shovels and pick axes in the removal of coal. Most of the coal seams of the Birmingham District were only a few feet thick, which made for back-breaking work. (BPL Collection.)

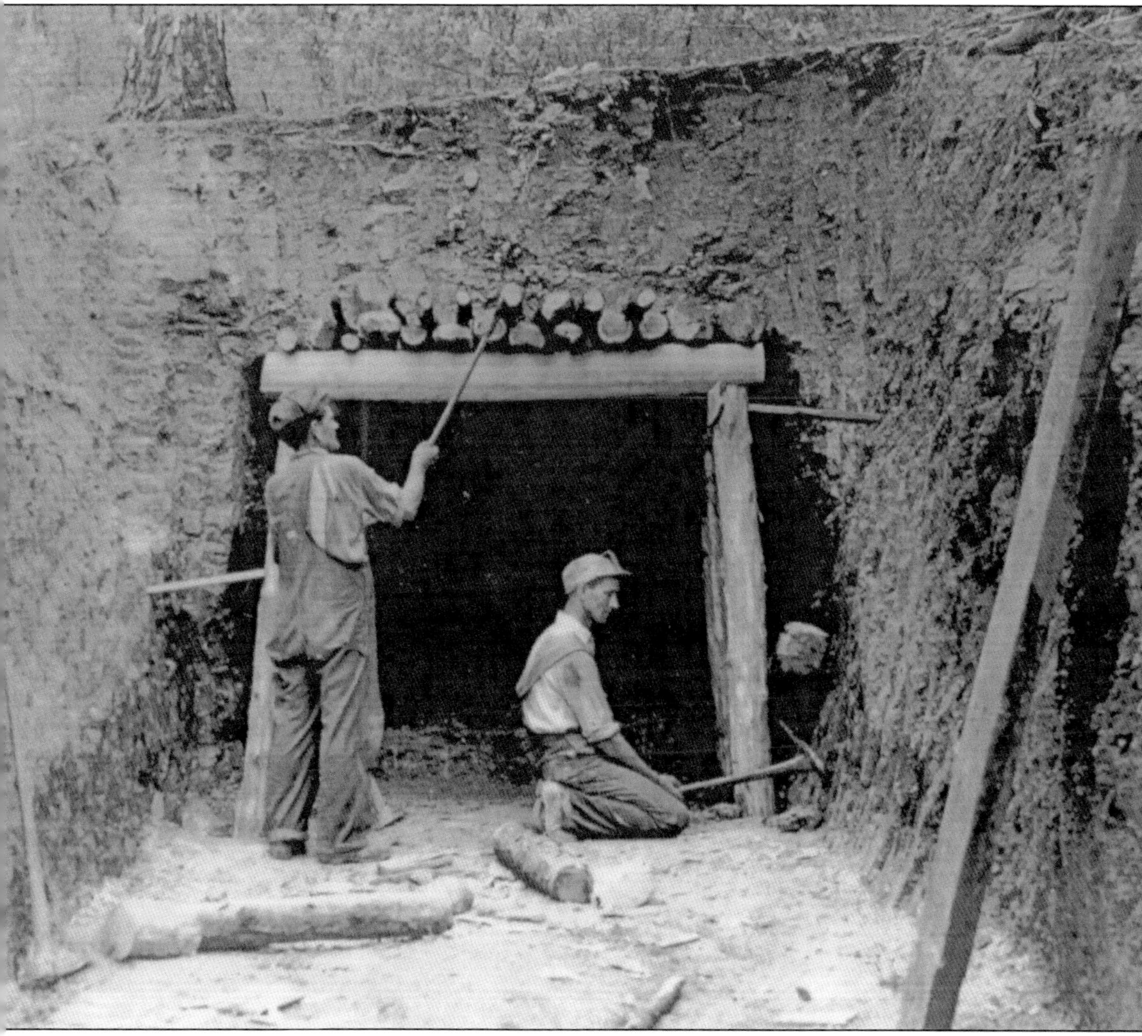

Many of the mines in the Birmingham district were drift mines that only required strong backs and manpower to open. The opening of mine faces were sometimes left to young boys. Wilbur Goode remembered that at age 13 he used picks and shovels to open mines in Mineral Springs, earning 15¢ a day. (BPL Collection.)

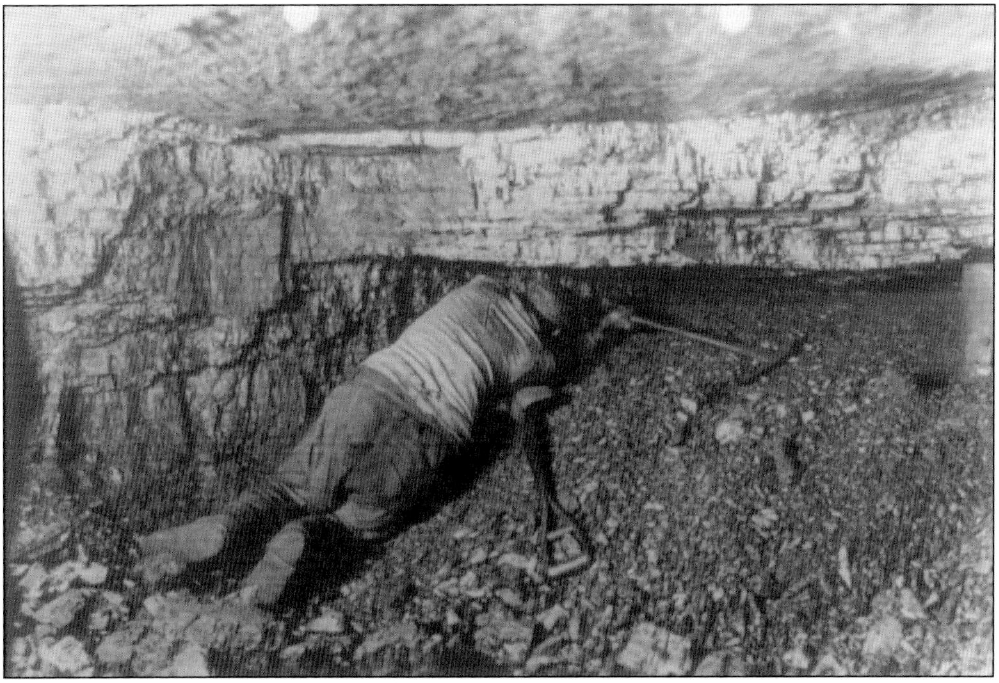

Many miners spent long hours on their backs and knees. This photograph is a representative example that shows the obvious dangers of the profession. (LOC.)

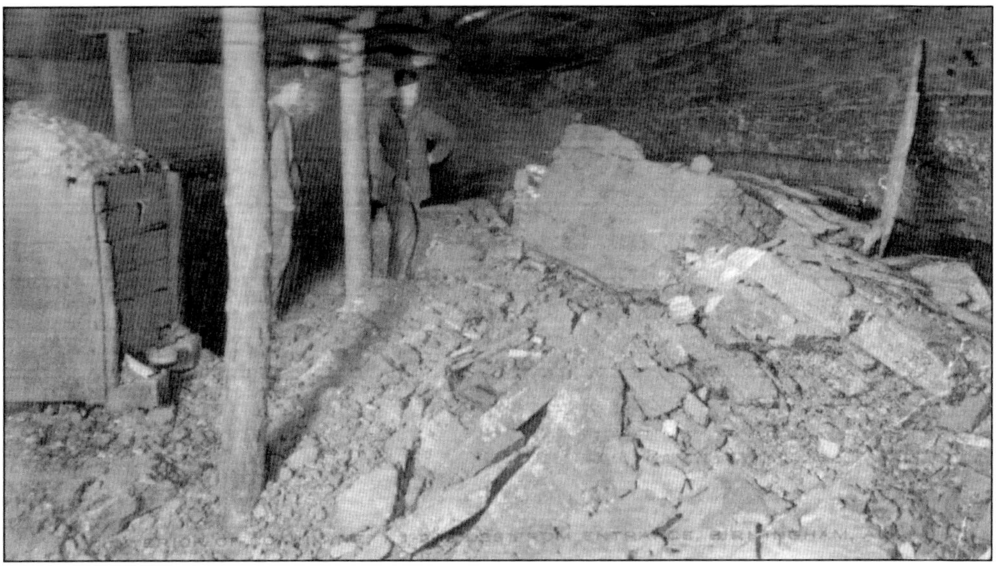

This early-1900s postcard reveals a scene 1.5 miles from the surface of a Birmingham-area mine. Notice the rough-hewn timber roof supports piles of debris. This coal mine contrasts sharply with the more mechanized mines of the district, such as those at Republic. (Author's collection.)

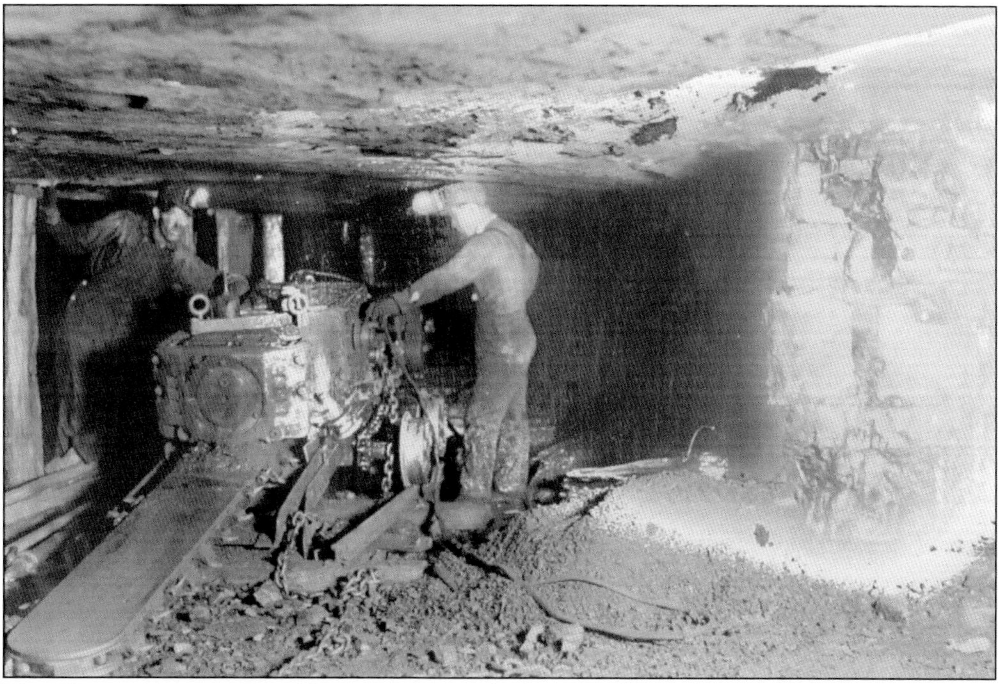

The Warner Mines at Republic were among the best mechanized in the area. This Library of Congress photograph is representative of the types of machinery used at Warner. (LOC.)

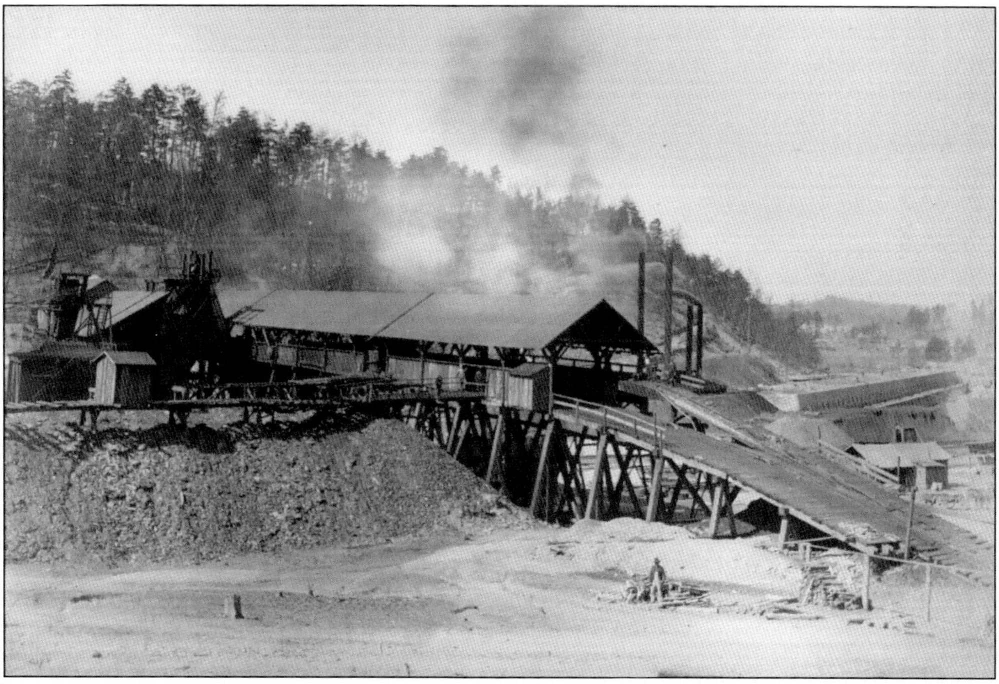

Mines in the Birmingham District were some of the most mechanized in the county. Consequently, industrial boosters recruited workers in European cities since many European mining operations were mechanized. The Warner Mines in Republic were some of the first to have electric trams. (BPL Collection.)

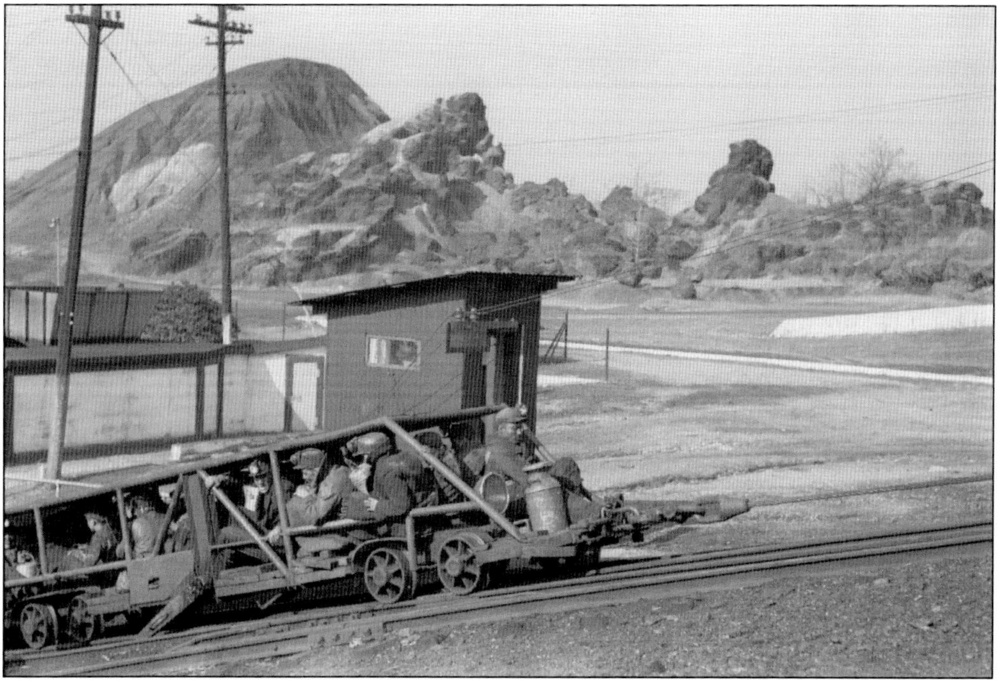

Many of the electric cars were driven by black employees. As the mines became more and more mechanized, the trams, originally made from wood, began to be made from iron. (LOC, FSA.)

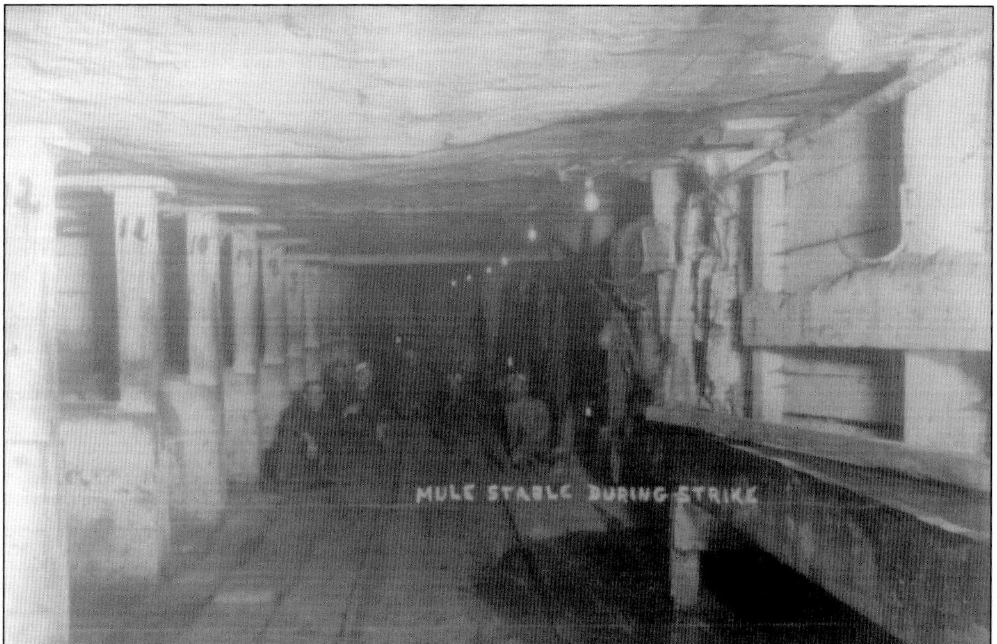

Sayreton mines, operated by Republic Steel, held the record for tons hauled in both a day and in a year. Mules used for the hauling were stabled underground on account of the depth. (Author's collection.)

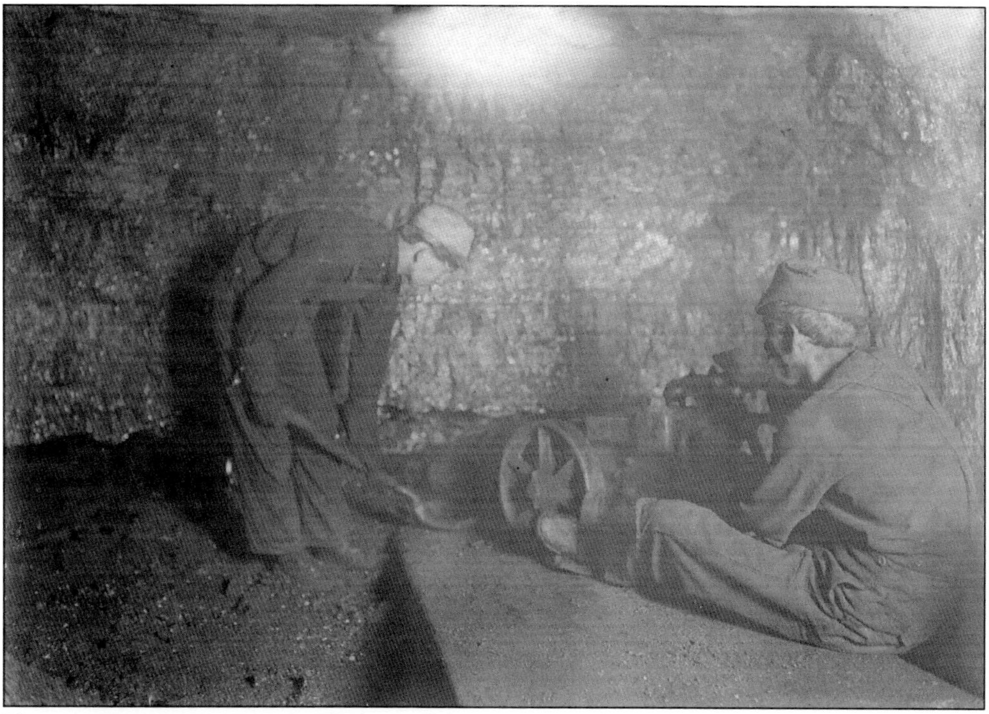

Electric equipment in mines made the work less taxing on those who had worked with picks, axes, and shovels but also those who labored pulling trams to the mouth of the mine. (LOC.)

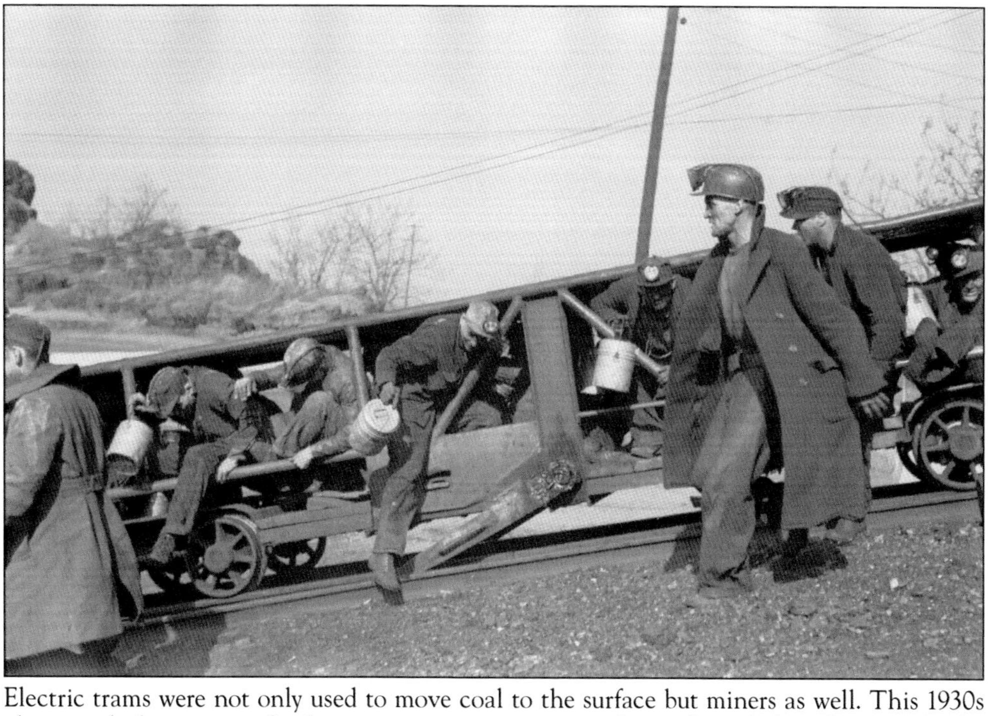

Electric trams were not only used to move coal to the surface but miners as well. This 1930s photograph shows a man-hauler tram transporting miners from a long shift underground. The tram would fill with the next shift for the return trip deep into the mine. (LOC, FSA.)

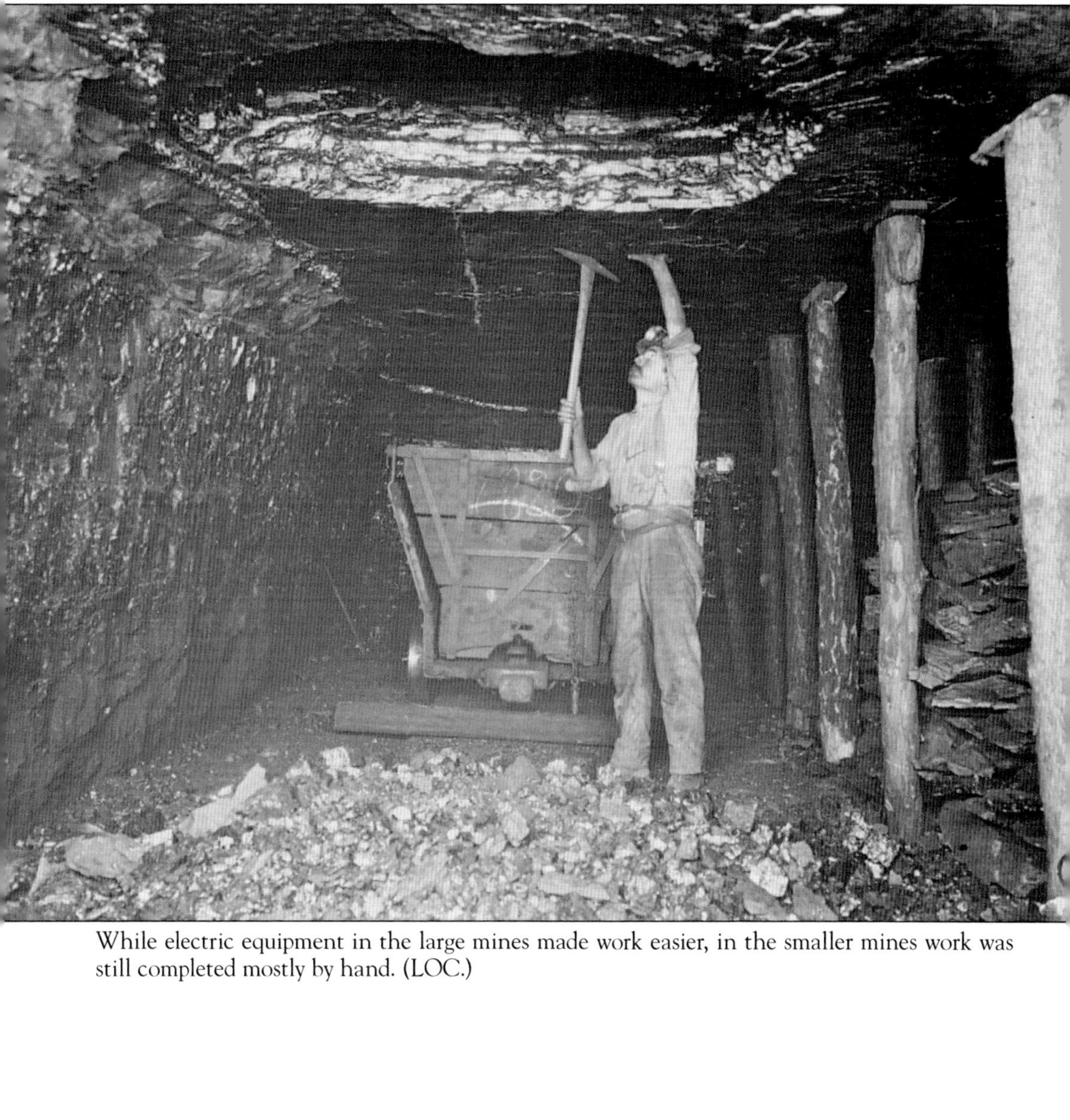

While electric equipment in the large mines made work easier, in the smaller mines work was still completed mostly by hand. (LOC.)

Two

MINERS

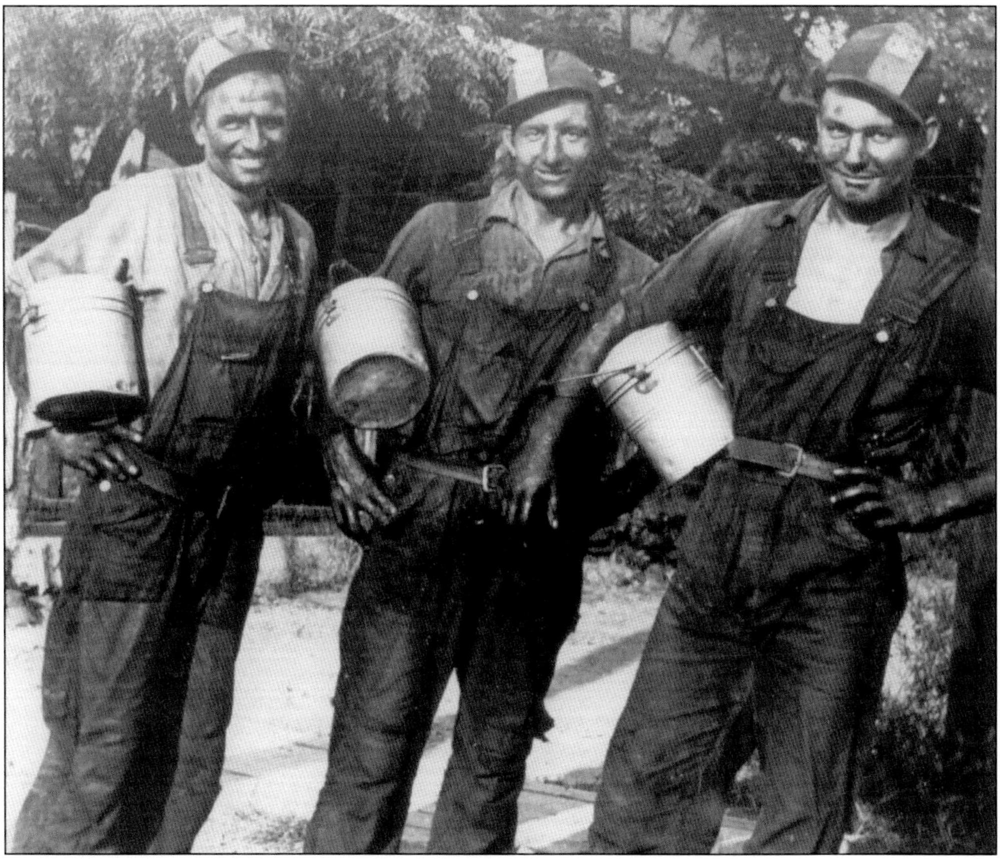

Many occupations have a standard uniform, and coal mining was no different. Overalls covered in coal, long sleeve shirt, boots, and soft cap with lantern made up the ensemble. When parts of the uniform became worn or tattered, wives repaired them, sometimes with unusual materials. Annie Sokira mended the knees of her husband's overalls with tire inner tube. (BPL Collection.)

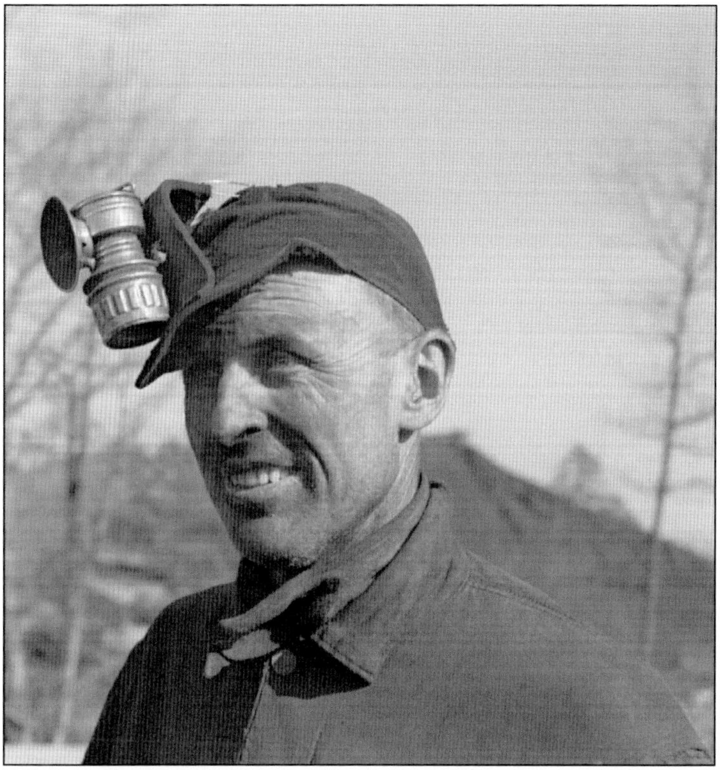

Before the advent of battery-powered lights, miners used carbide lamps. The lamp has two sections. The bottom was filled with carbide and the top reservoir with water. The mixture of the two lead to a flame that burned for hours without needing to be refilled. (LOC, FSA.)

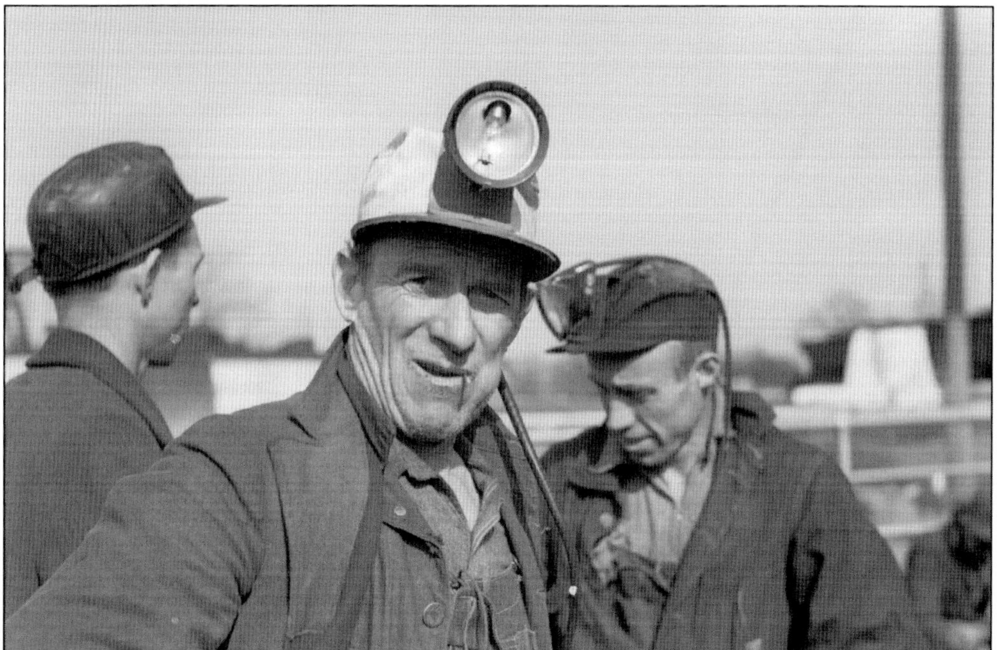

After battery-powered lamps came onto the market in the 1930s, the carbide lamps were no longer used. A miner wore a battery attached to his belt with a cable running up to the headlamp. (LOC, FSA.)

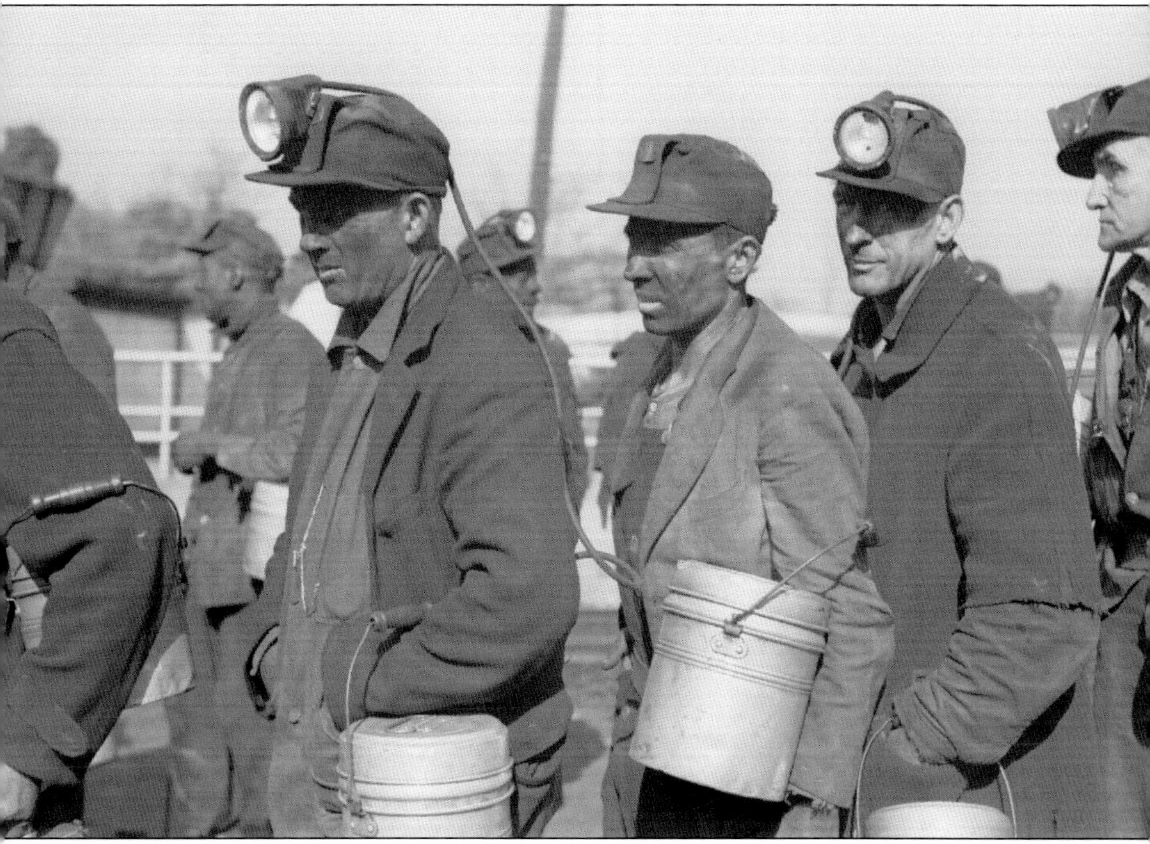

Born in Bisacquino, Sicily, Barto Tombrello came into the country and made his way to Cardiff. His first attempt to work in the mines was unsuccessful, as he appeared to be underage. Determined to work, he jumped into line, got tools, and began his first job making 90¢ for a 10-hour day. (LOC, FSA.)

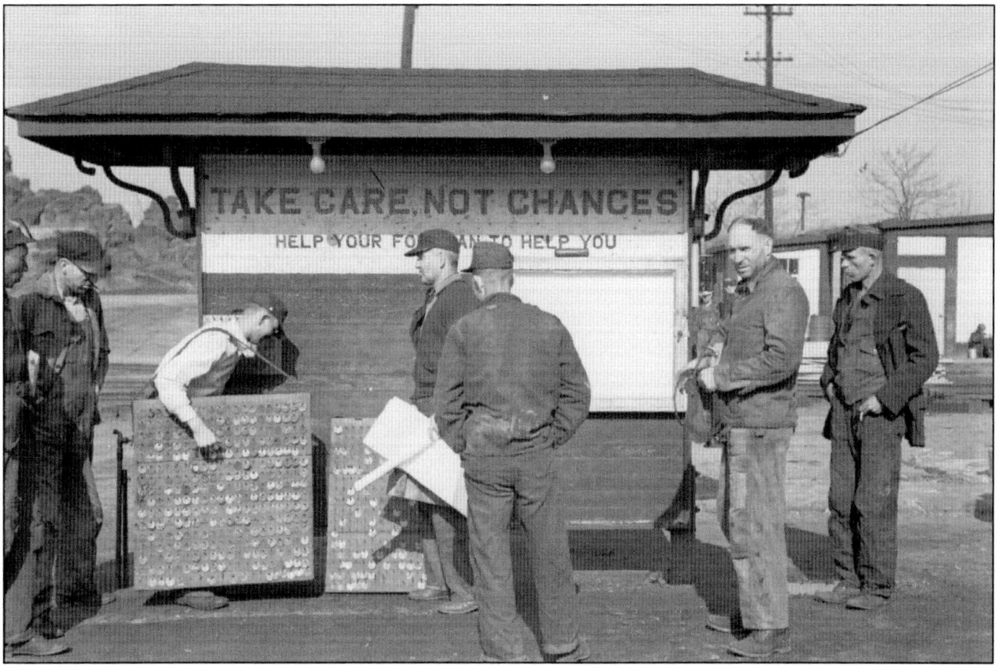

These miners stand in line to pick up their coal car check tokens. These tokens were imprinted with a number that was assigned to each miner. Once the miner had completed his quota of coal in the car he was assigned, he attached the token to the car and sent the car to the surface. He would be credited with the coal that he had mined. (LOC, FSA.)

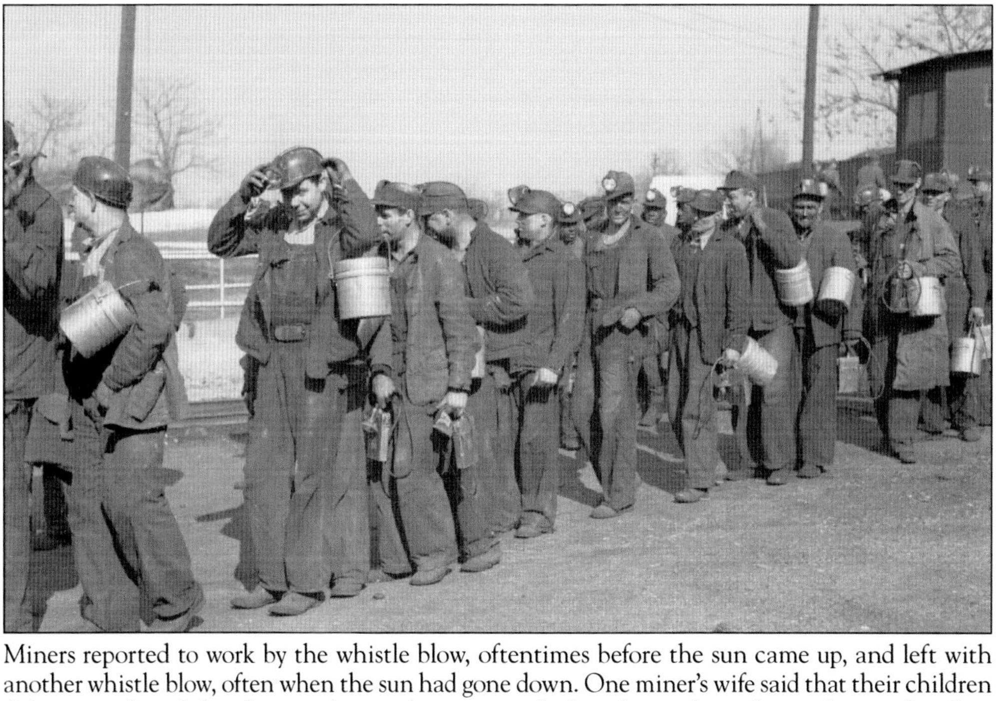

Miners reported to work by the whistle blow, oftentimes before the sun came up, and left with another whistle blow, often when the sun had gone down. One miner's wife said that their children did not see their father for a week since he was gone before they woke and came home after they were asleep. (LOC, FSA.)

This miner waits on his shift to begin. (LOC, FSA.)

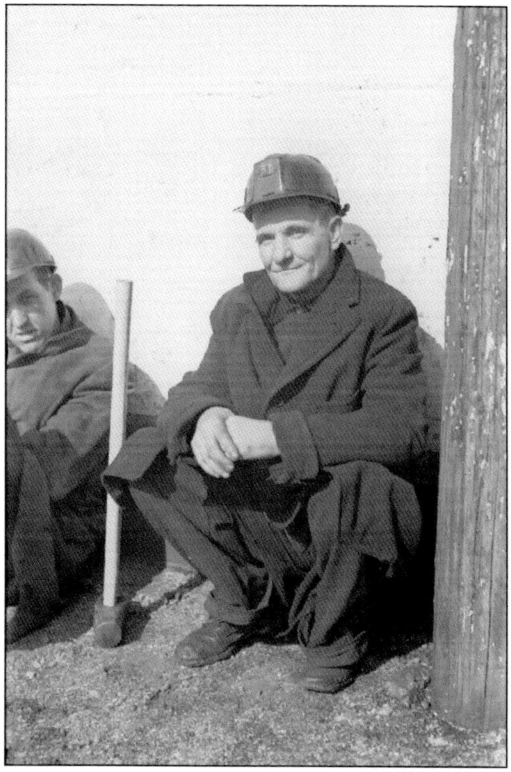

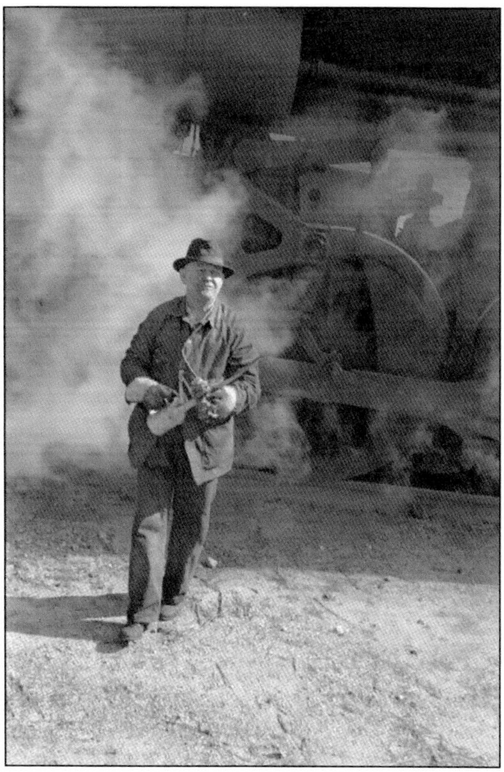

Here, a greaser on a coal yard rail line keeps everything in working order. Young boys at the mine had this job on coal cars. (LOC, FSA.)

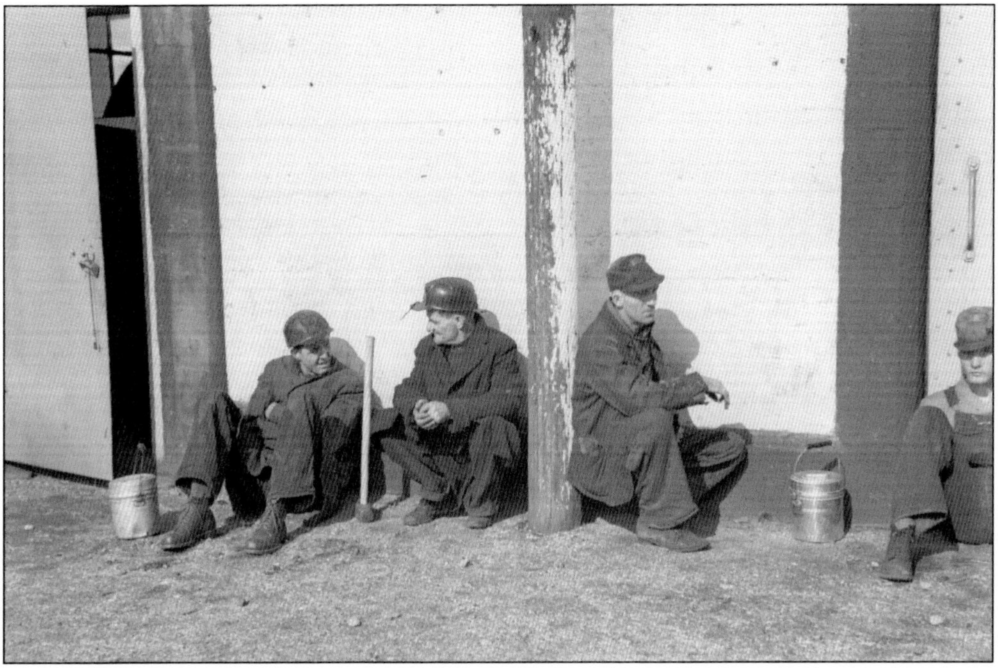

These miners patiently wait for a shift change. (LOC, FSA.)

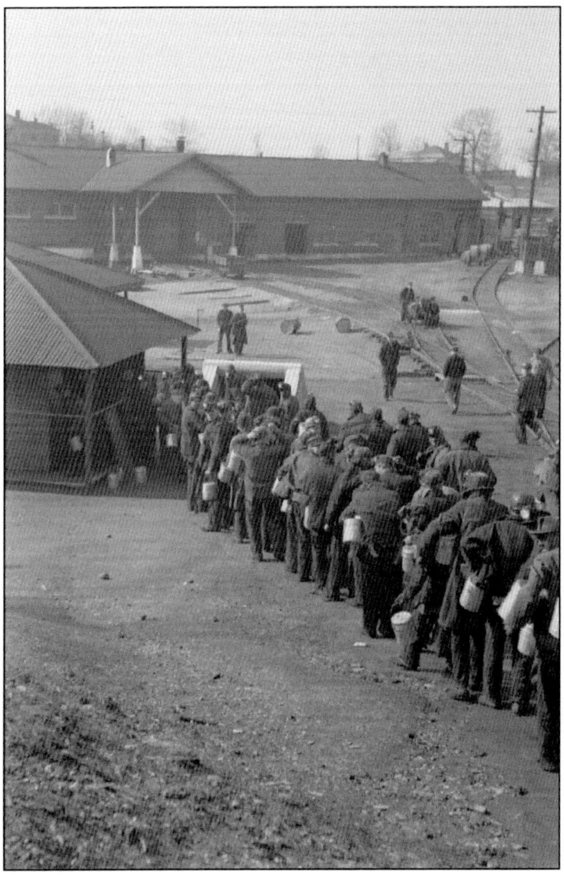

These miners wait in line for the tram that will carry them thousands of feet underground. (LOC, FSA.)

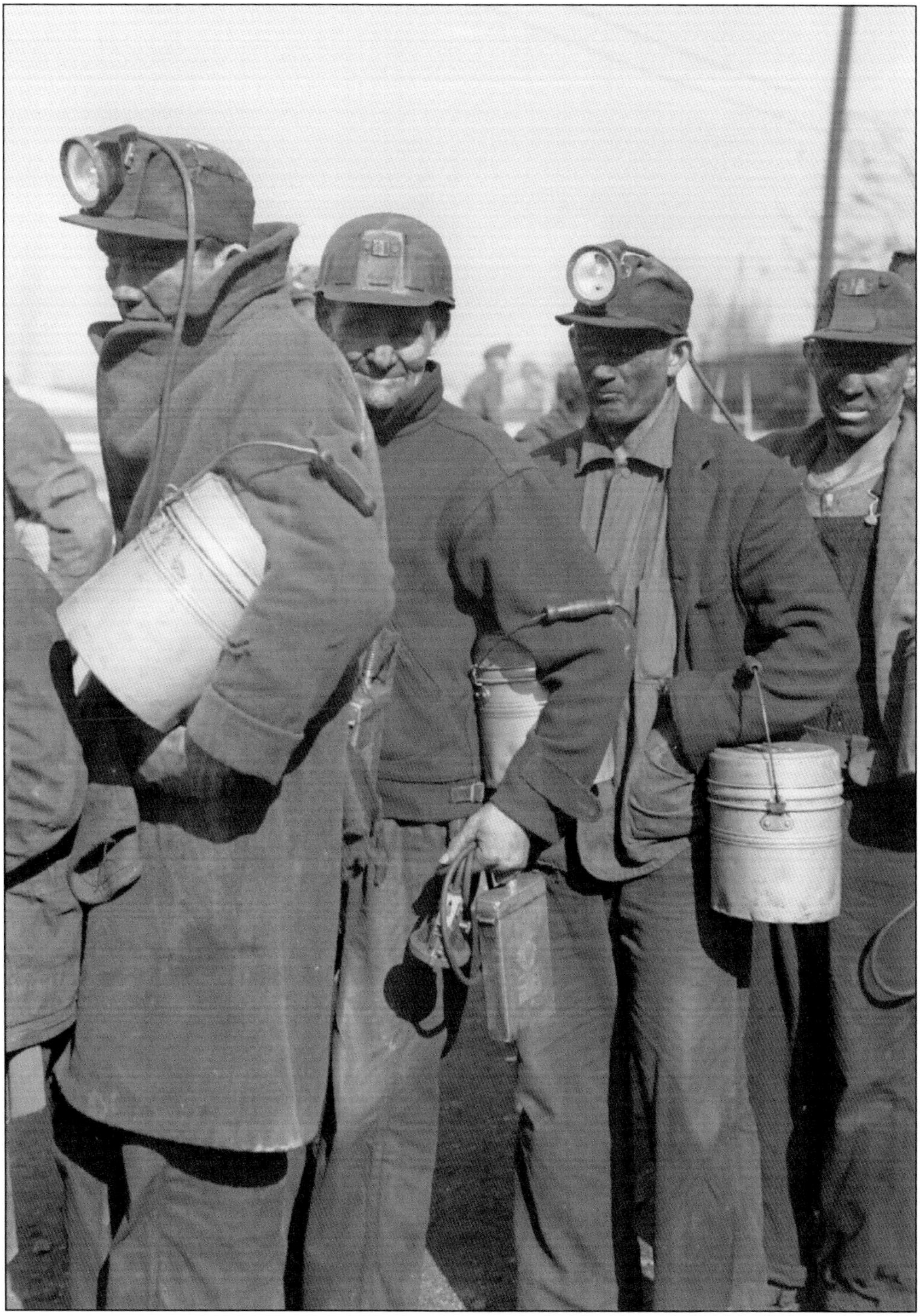

One of the staples of the miner's gear was his lunch pail. Lunch pails were divided into two or three sections. The bottom held water or, for some, their ration of liquor. The top portion held a sandwich or their lunch contents. The metal bucket served a purpose too. It kept mice out while in the dark confines of the mine. (LOC, FSA.)

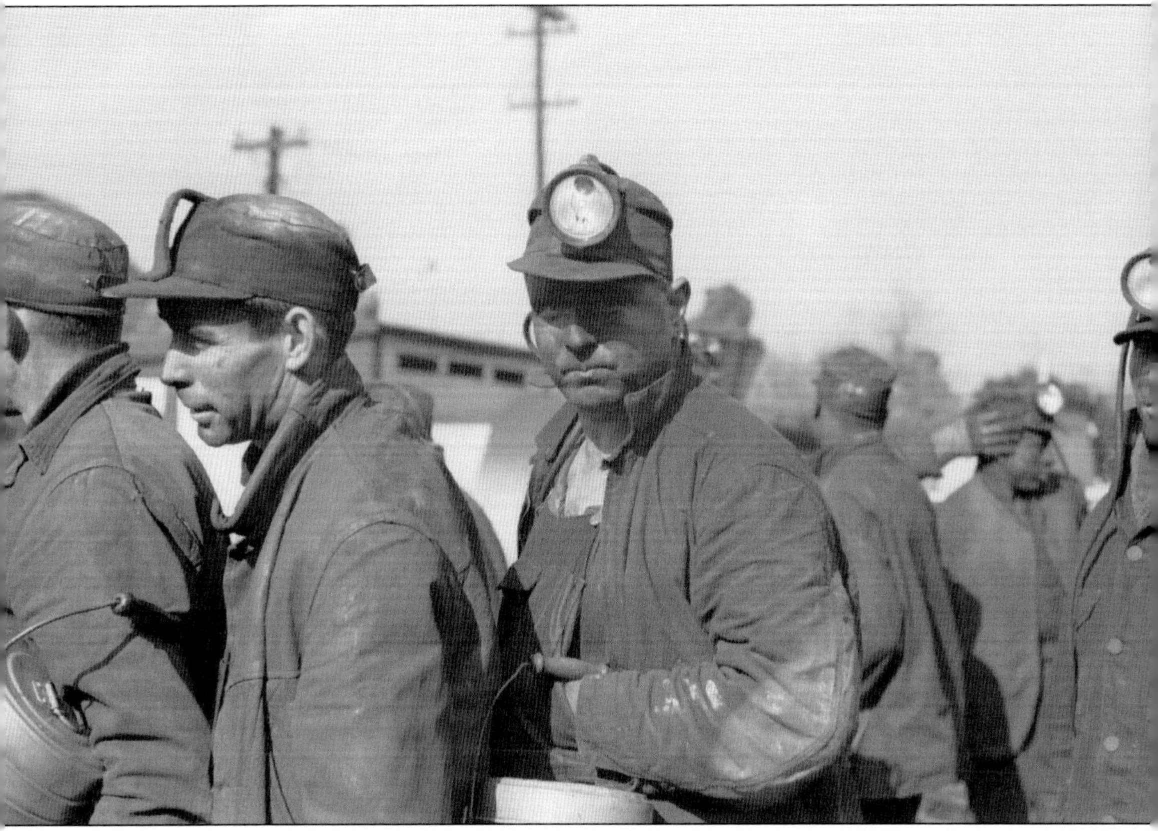

This miner's face shows the fatigue that many miners endured. Many worked on their stomachs in mines with seams only a few feet thick. They constantly faced threats from cave-ins, methane gas and explosion, and damp conditions. Even with these dangers, they also faced the potential for wage cuts, strikes, and mine closure. (LOC, FSA.)

Convict labor was a shameful compact developed between coal operators and state and local prison systems. Prisoners, mostly black, found themselves chained with leg irons and pressed into service and paid a small wage while the coal operators and prisons profited from this form of industrial slavery. These miners toiled at Banner Mines. (BPL Collection.)

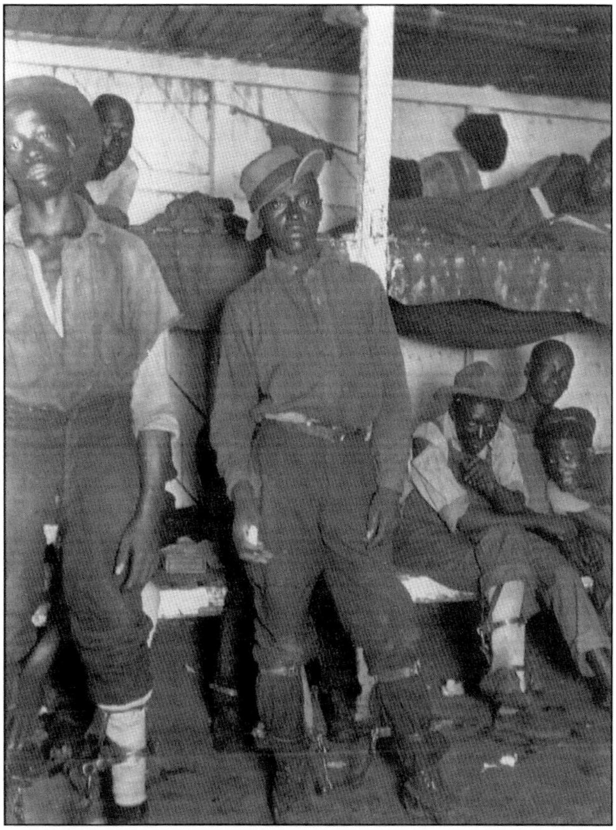

Banner Mines near Graysville blew in 1911, killing some 128 miners, 123 of whom were black. Explosions were not uncommon. Coalburg blew in 1890, Overton in 1925, and many others in between. One miner's daughter from Brookside remembered when a mine exploded that she was told her father had died. Crying, she happened to look up and saw her father approaching covered in coal dust. (BPL Collection.)

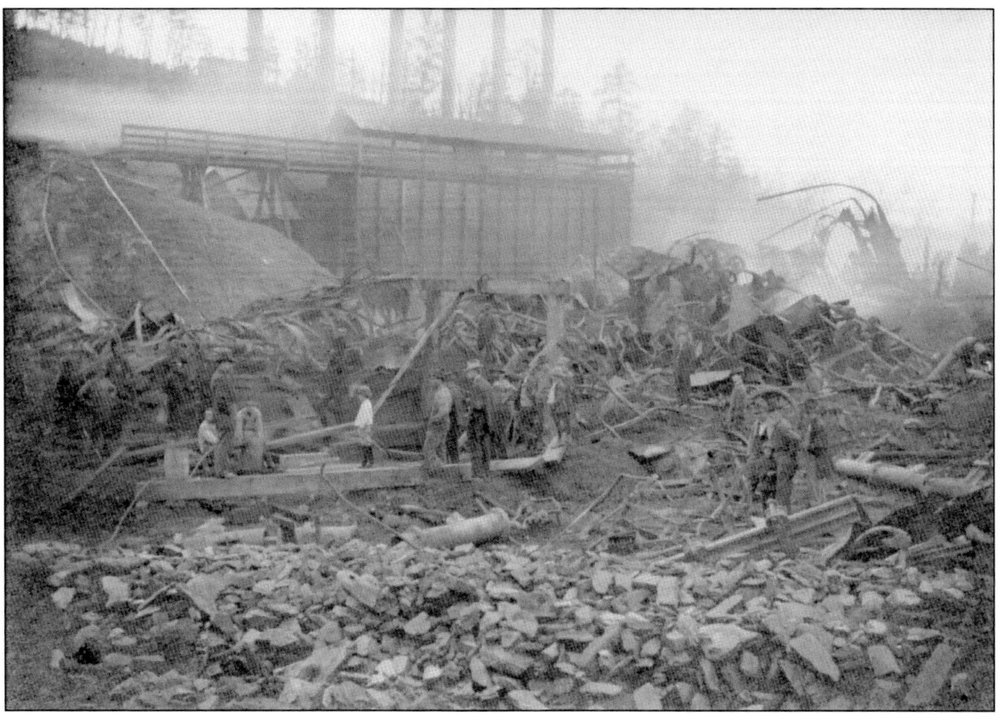

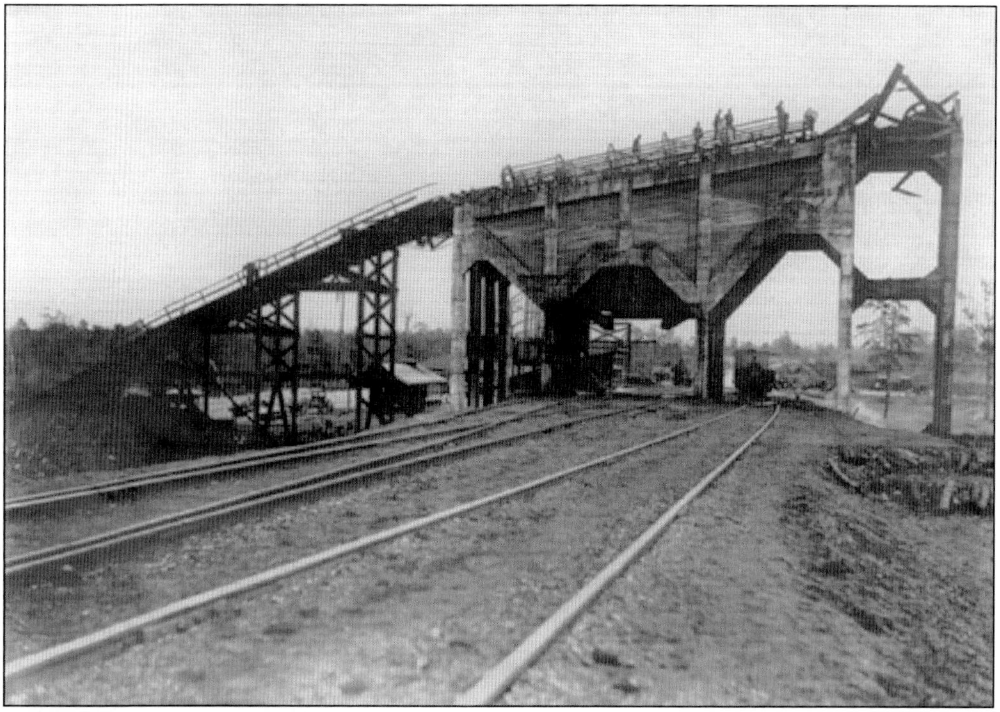

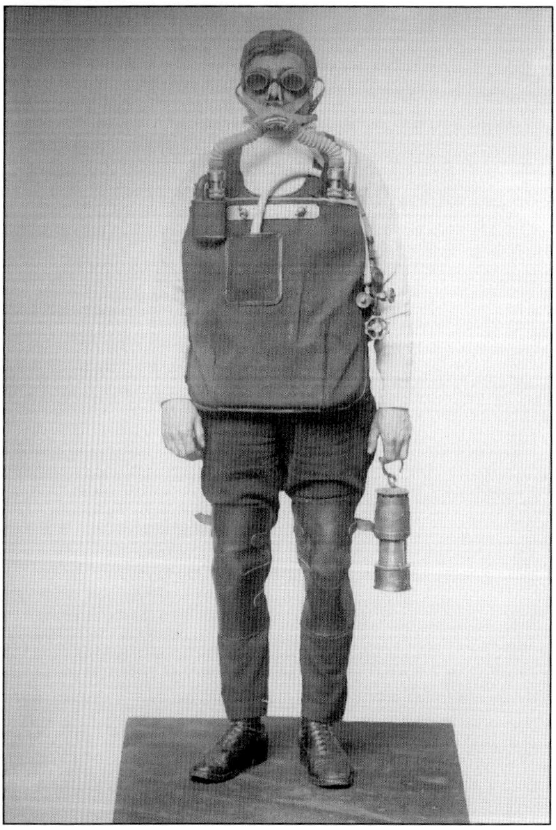

Dolomite Mine No. 3 exploded in November 1922, trapping some 475 miners. Officials reported that three transport cars careened 1,100 feet down the slope to the mine below. Newspapers detailed that the cars cut an electric current that sparked the fire. The fire burned all the way to the main tipple. (LOC, FSA.)

When Edgewater Mine blew on July 30, 1948, eleven miners were killed and over 600 TCI (Tennessee Coal Iron and Railroad Company) miners refused to enter three of the company's other mines. When many were afraid to enter a mine, rescue teams took over. This is an example of a mine rescuer fully outfitted (LOC.)

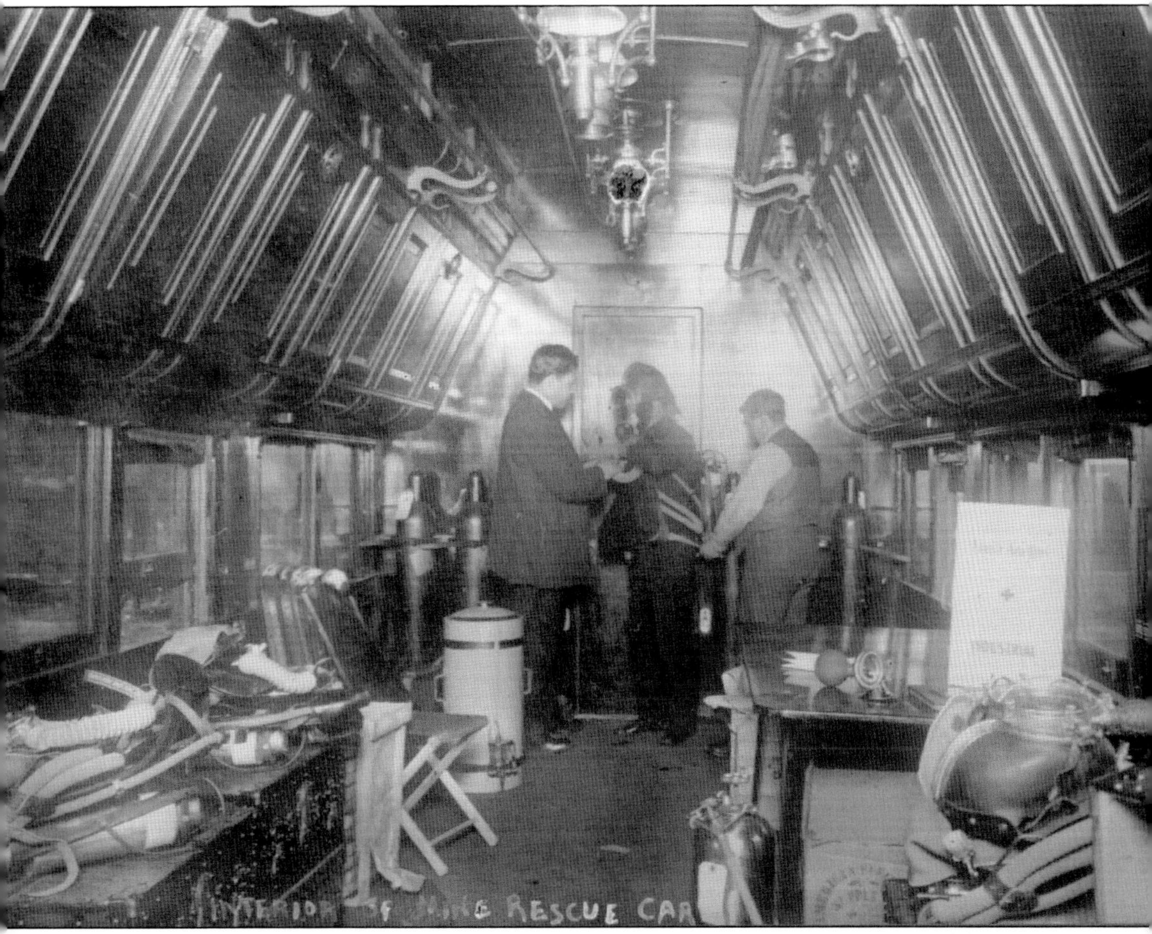

When the Palos Mine blew in May 1910, the scene was horrific. Rescue teams from Birmingham came in by train to assist in the recovery efforts. This is an example of a mine rescue recovery tram. So devastating was this disaster, William Glass wrote a song, "The Palos Mine Disaster." (LOC.)

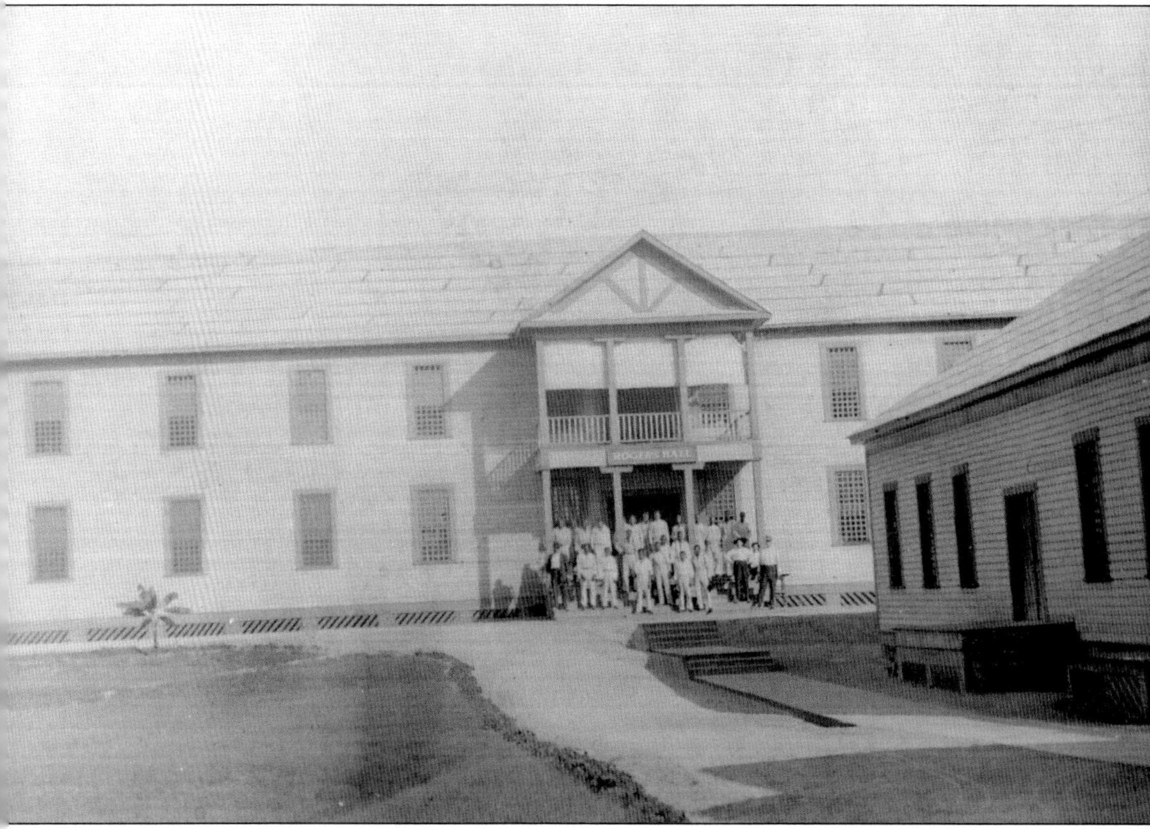

While this photograph portrays a model prison camp complete with dormitories, mess hall, kitchen facilities, and hospital, reality was much different. Prison camps were far from this ideal. They were dark, dirty, and confining. Many prisoners suffered from disease and threats of beatings if they disobeyed or threatened escape. (BPL Collection.)

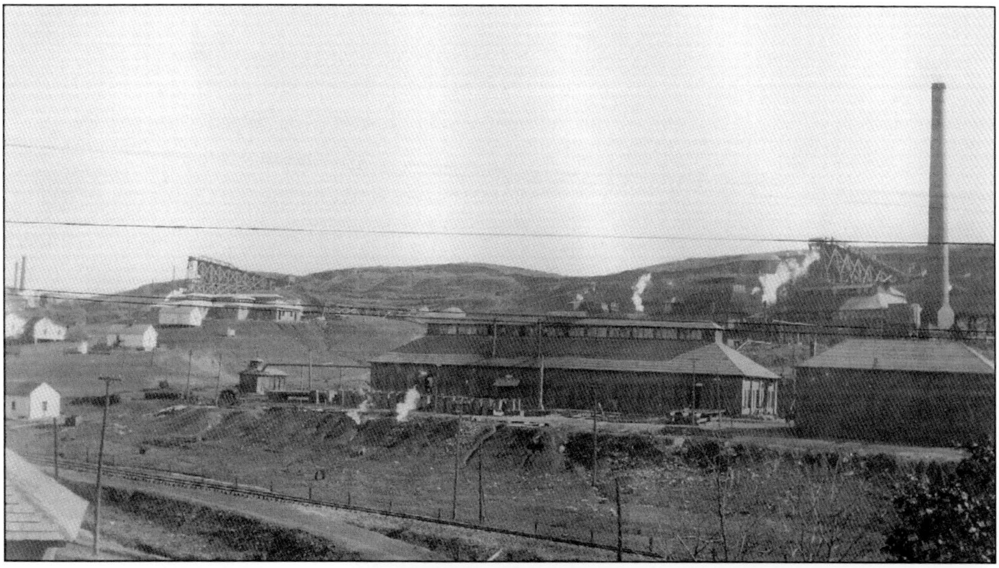

Tennessee Coal Iron and Railroad Company (TCI) became the largest player in the steel industry in Birmingham and one of the largest owners of coal mines and coke producing facilities. Moreover, TCI was the largest contractor in the convict lease system. (LOC.)

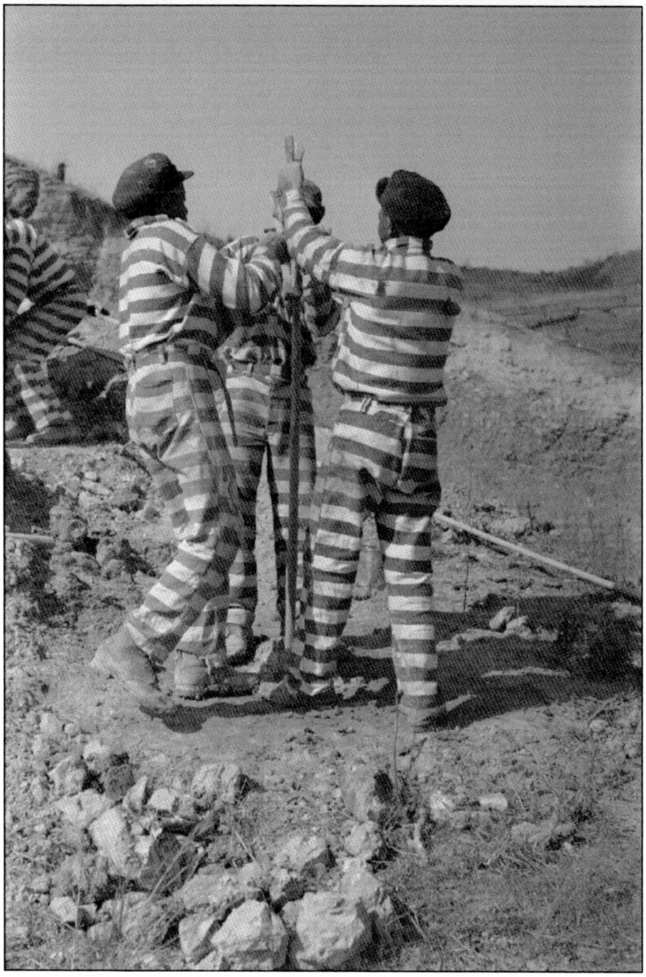

Being sentenced to a prison camp proved one of the worst fates that an alleged criminal could receive. Prisoners were shackled at all times. Some prisoners were hurried to the mines through a back door leading directly to the mines. The public was never aware of the true nature of prison camp life. (BPL Collection.)

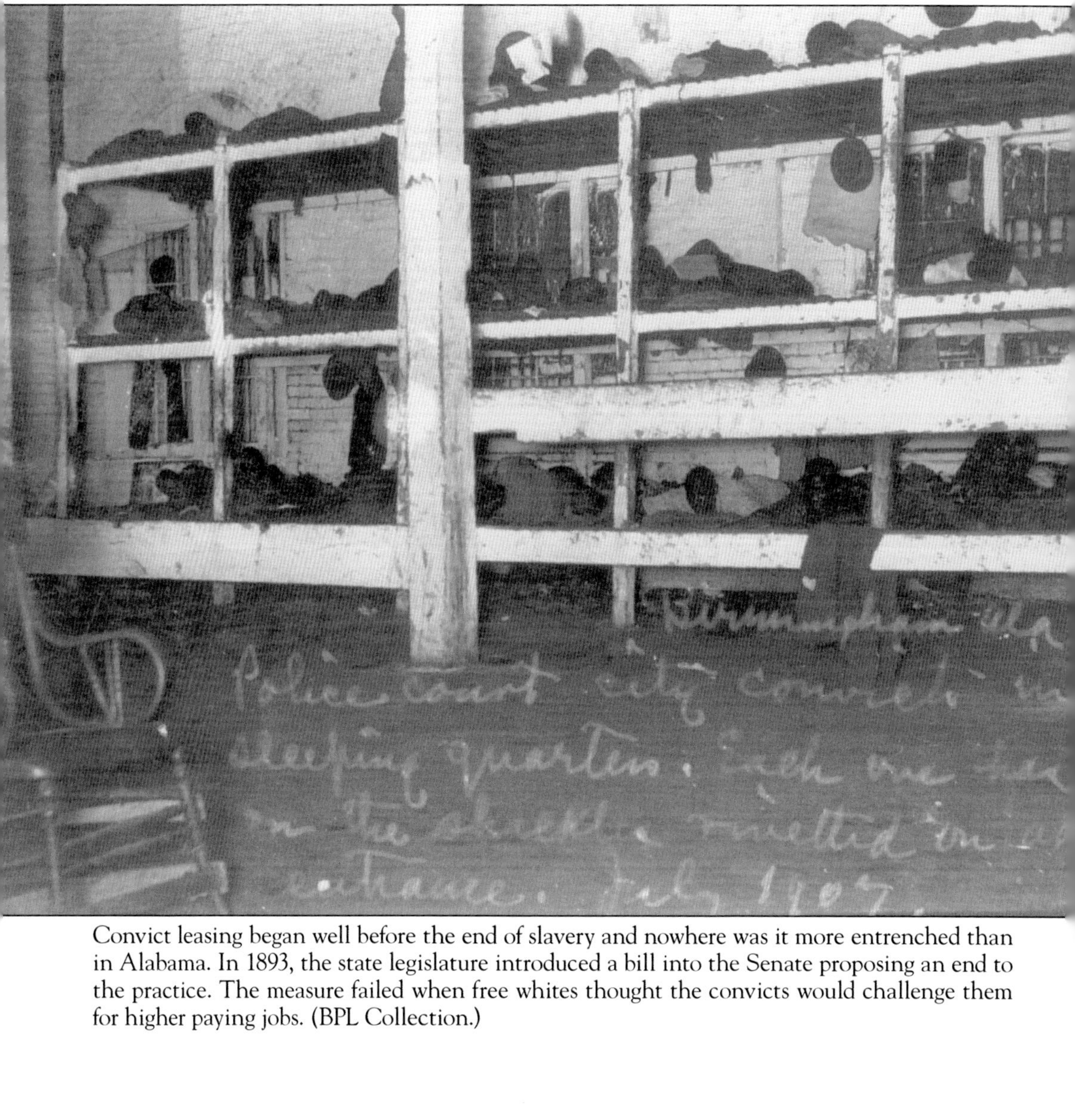

Convict leasing began well before the end of slavery and nowhere was it more entrenched than in Alabama. In 1893, the state legislature introduced a bill into the Senate proposing an end to the practice. The measure failed when free whites thought the convicts would challenge them for higher paying jobs. (BPL Collection.)

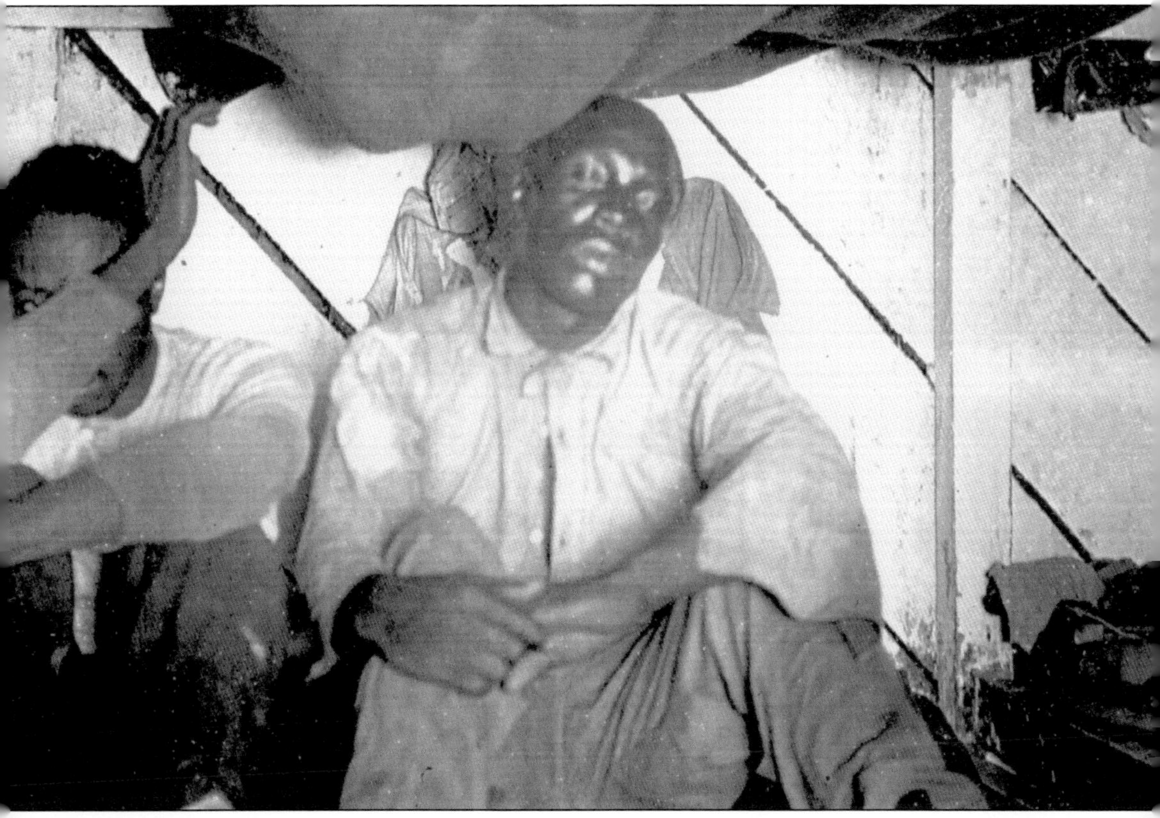

During the days of Jim Crow, the city of Birmingham leased hundreds of convicts to mine coal. By 1900, the majority of coal miners at some mines were convicts. Many were convicted of harassing white women, gambling, and vagrancy. In the system, they suffered dysentery, tuberculosis, whippings, and hangings. (BPL Collection.)

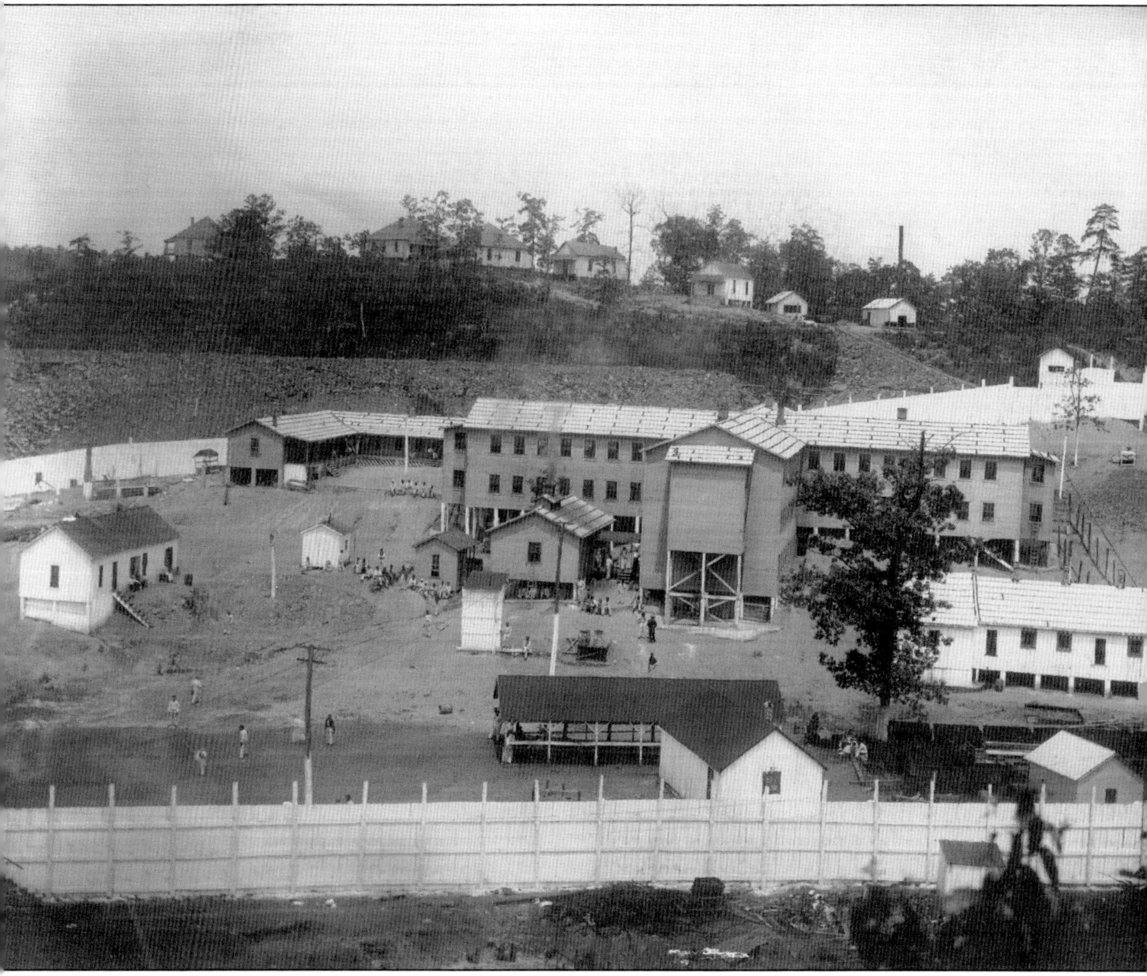

One camp doctor wrote in the *Independent* in 1907 that camp prisons were rife with drug abuse, gambling, and a litany of offenses large and small. The doctor, who wrote anonymously, observed convicts playing cards and rolling dice directly under the noses of prison guards. Moreover, he argued that this love of gaming followed many into the free world. (BPL Collection.)

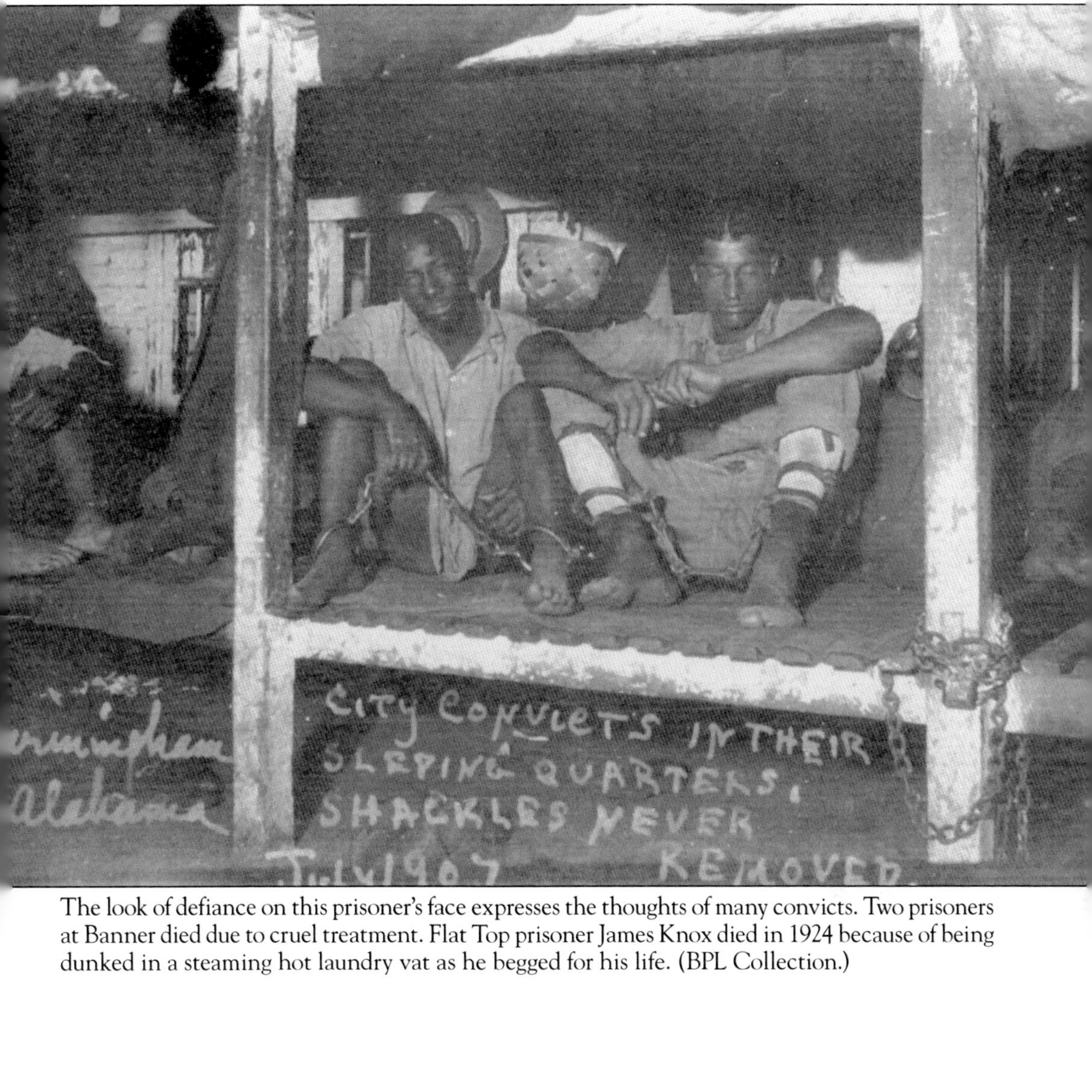

The look of defiance on this prisoner's face expresses the thoughts of many convicts. Two prisoners at Banner died due to cruel treatment. Flat Top prisoner James Knox died in 1924 because of being dunked in a steaming hot laundry vat as he begged for his life. (BPL Collection.)

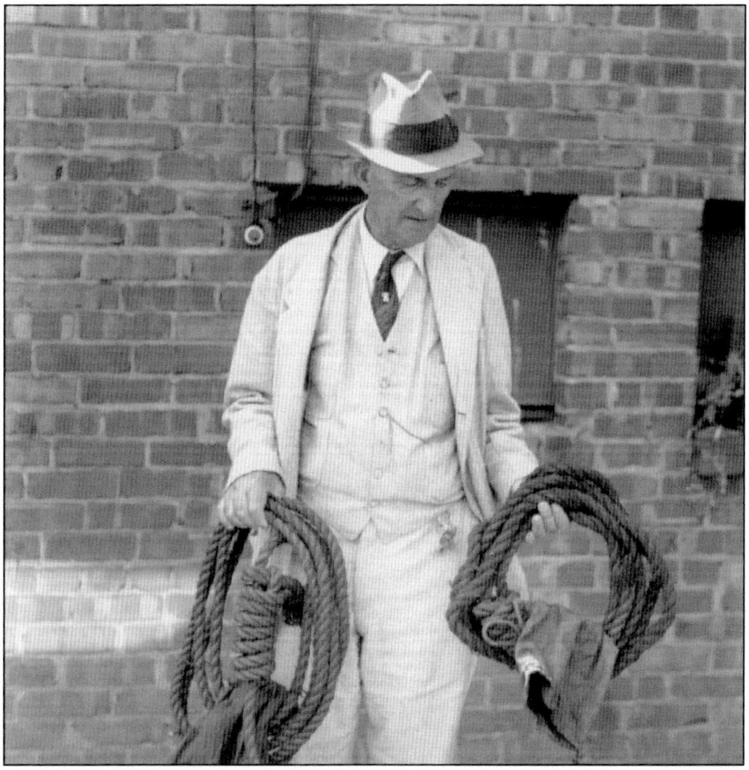

The convict lease system was not merely a system for the enrichment of coal operators and prison systems, but a means of subjugation for blacks and poor whites as well. The Alliance prison where this photograph was taken was the place of hangings and whippings for convicts before the system was done away with in the 1930s. (BPL Collection.)

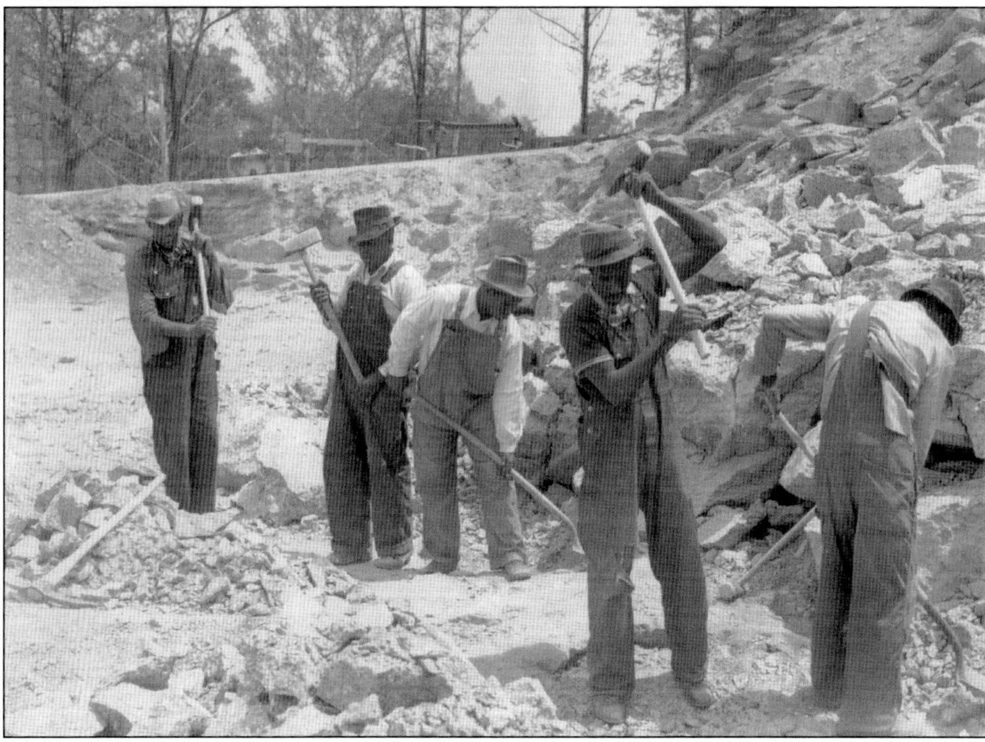

This was a typical day's work at a convict labor camp. (BPL Collection.)

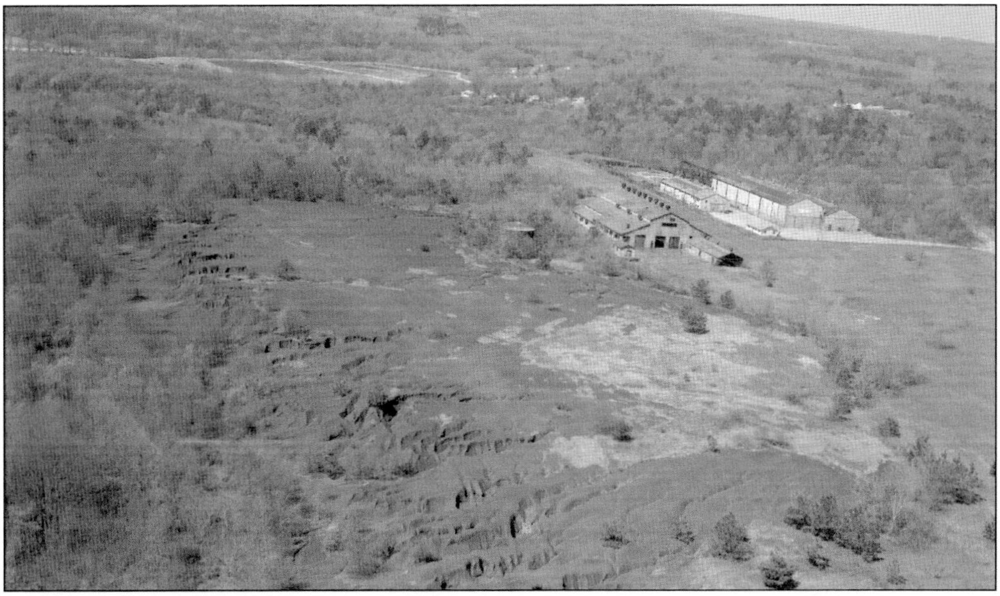

Convicts were thought of as industrial slaves, with little thought given to their care except to provide enough sustenance to allow them to continue working. When they died, many were hastily buried. In Jefferson County, many of their graves were unmarked and are now overgrown, like in this Pratt Coal and Coke Convict Cemetery. (LOC, HABS.)

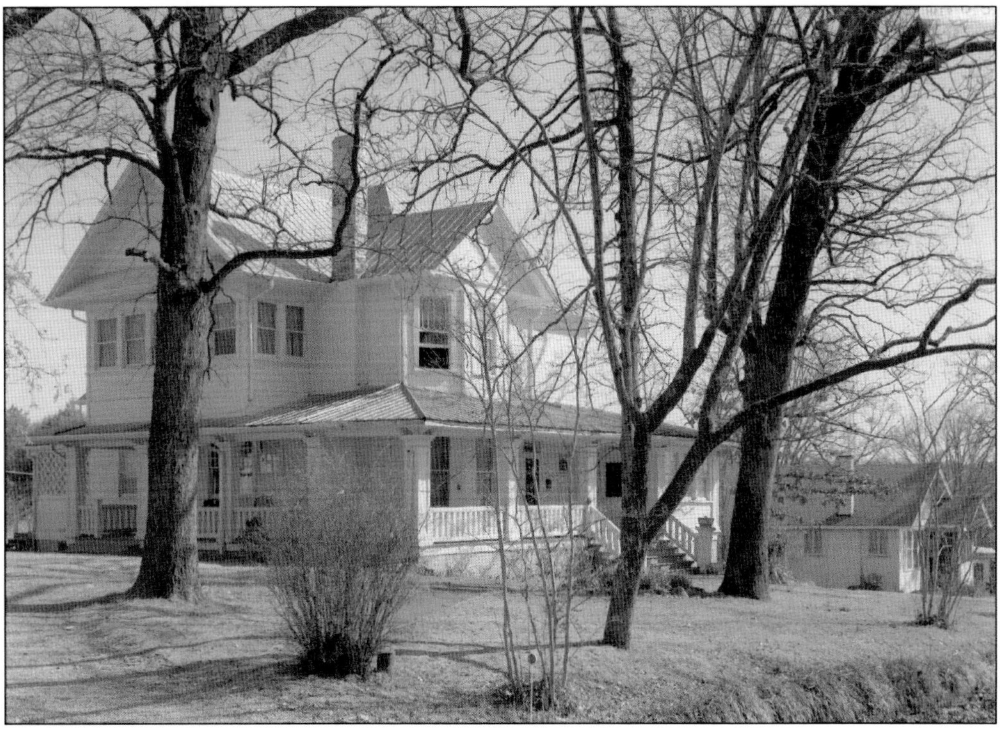

This house, owned by TCI president Don Baker (1901–1907), stands in stark contrast to the scenes of convict leasing on the previous pages. It would be George Gordon Crawford who would end TCI's affiliation with convict leasing in 1911. (LOC, HABS.)

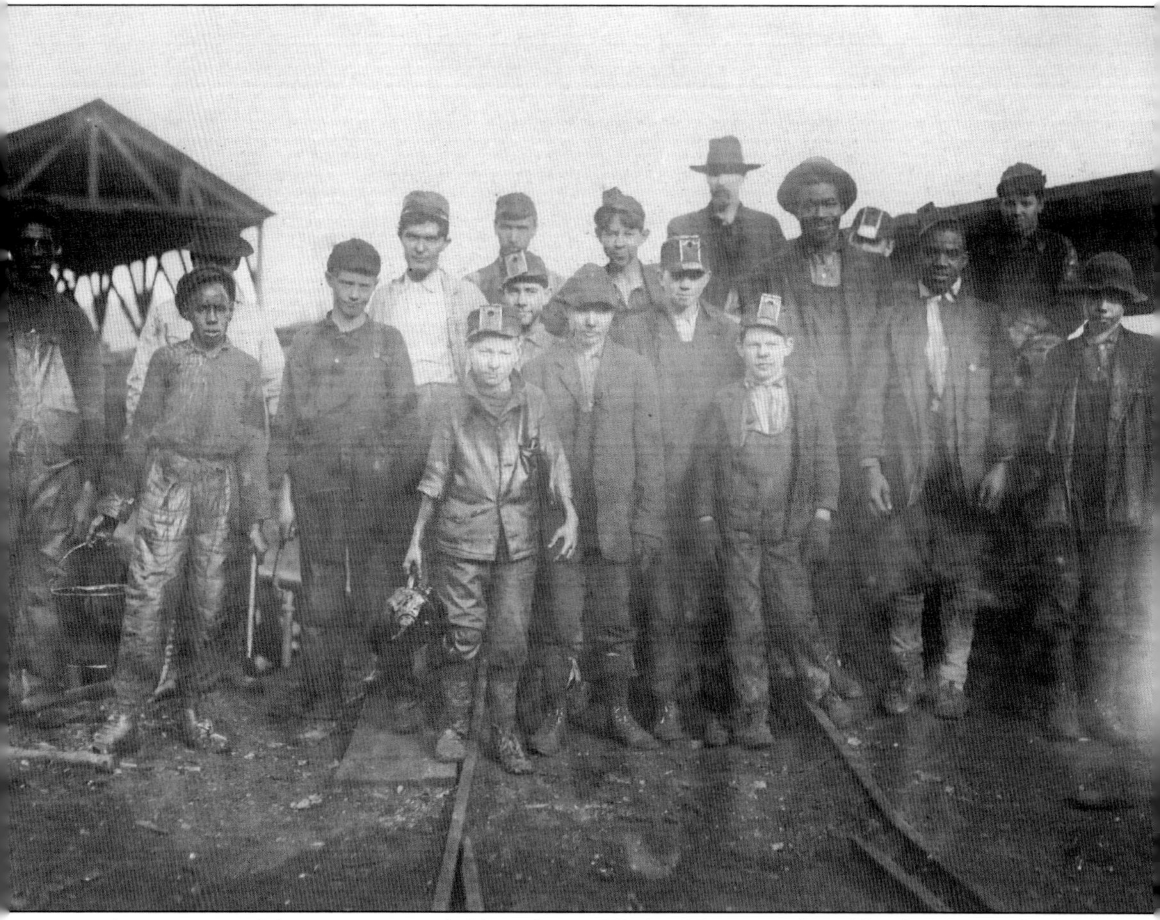

An important labor source was children. Lewis Wickes Hines was hired by the National Child Labor Committee to document the atrocities of child labor throughout the country. In Alabama, he focused on the textile and coal industries. His photographs lead to a congressional investigation and new child labor laws. (LOC, NCLC.)

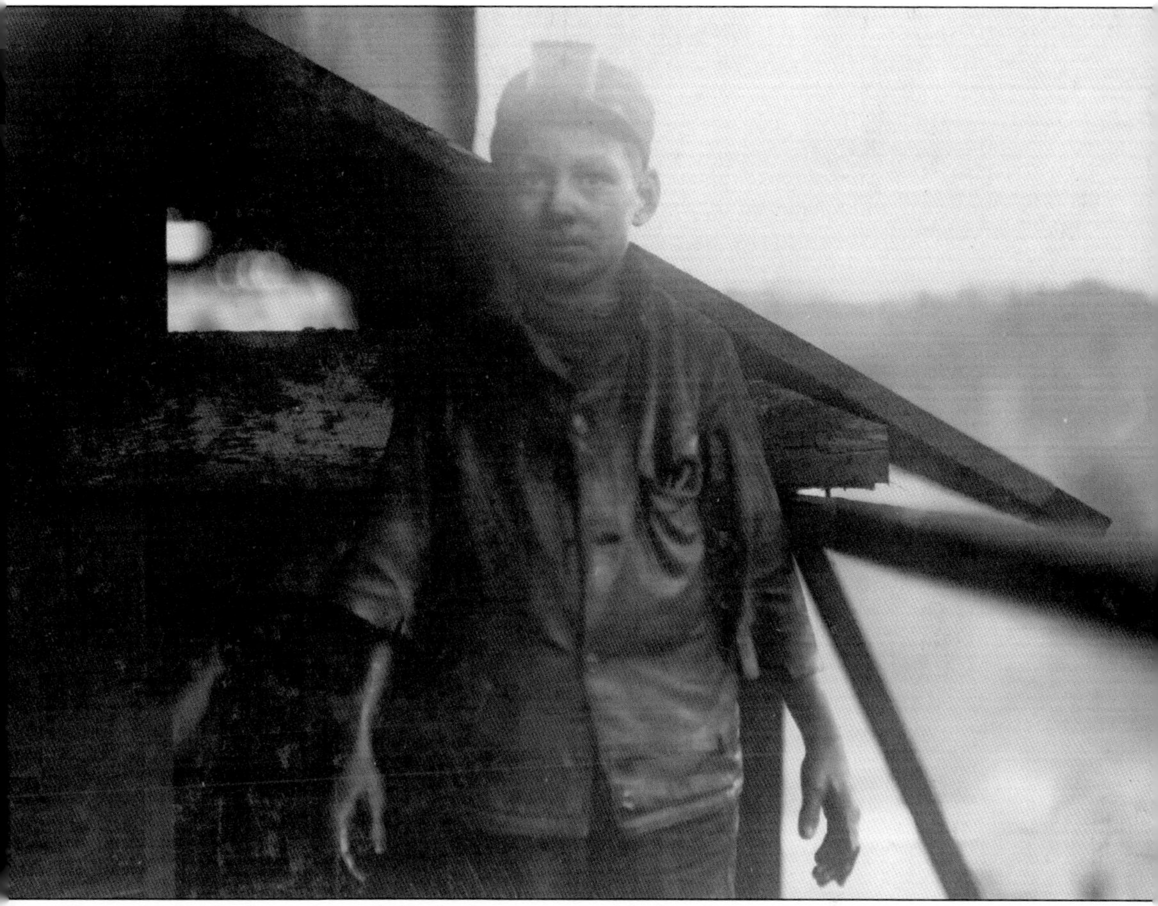

Shorpy Higginbotham claimed that he was 12 years old when Lewis Hines took this photograph in 1910, though Hines doubted Shorpy's claims. Shorpy was a "greaser" who kept all moving parts on tramcars oiled and functioning, which required carrying heavy buckets of grease around the yard. (LOC, NCLC.)

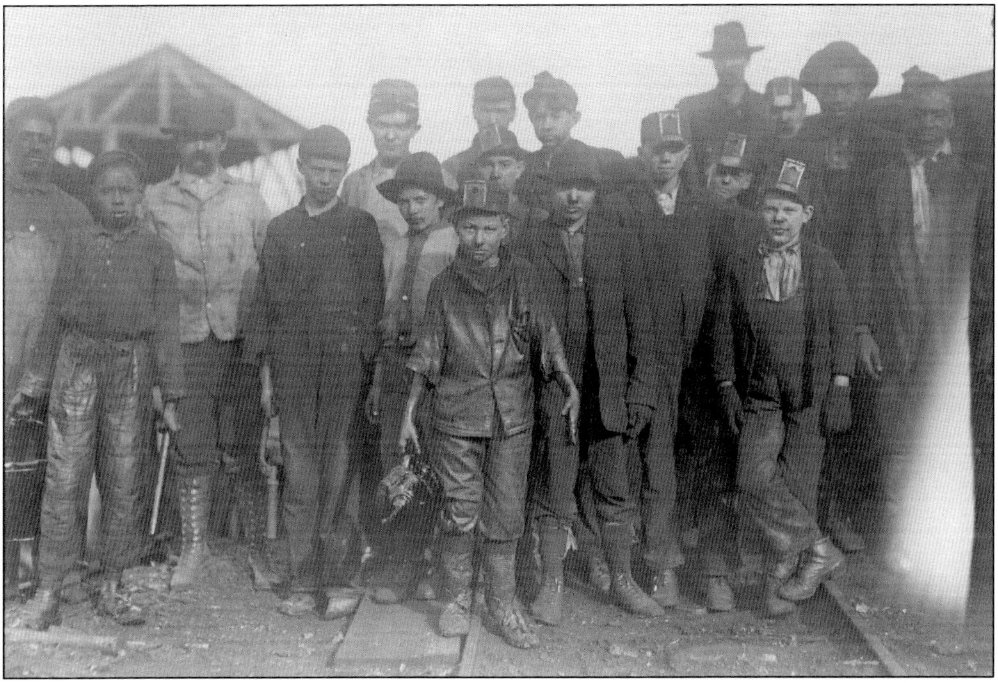

In the late 1800s, miners at Coketon struck, protesting their paying of "pushers" to move coal cars out of the mines. They argued that this was the company's responsibility to pay pushers. The company's solution: hire young boys at a reduced wage. (LOC, NCLC.)

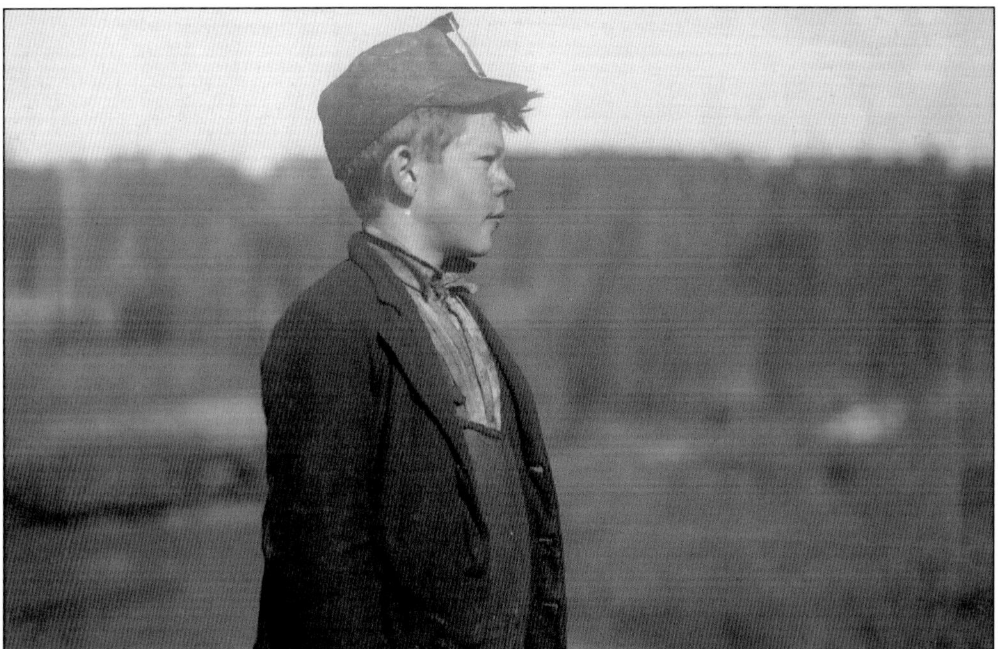

Dave was a young pusher who claimed to be 14 years old, though this photograph suggests that he may have been quite a bit younger. Dave had one of the most dangerous jobs in the mines—moving tramcars around. Many of the young pushers were between 9 and 12 years old. (LOC, NCLC.)

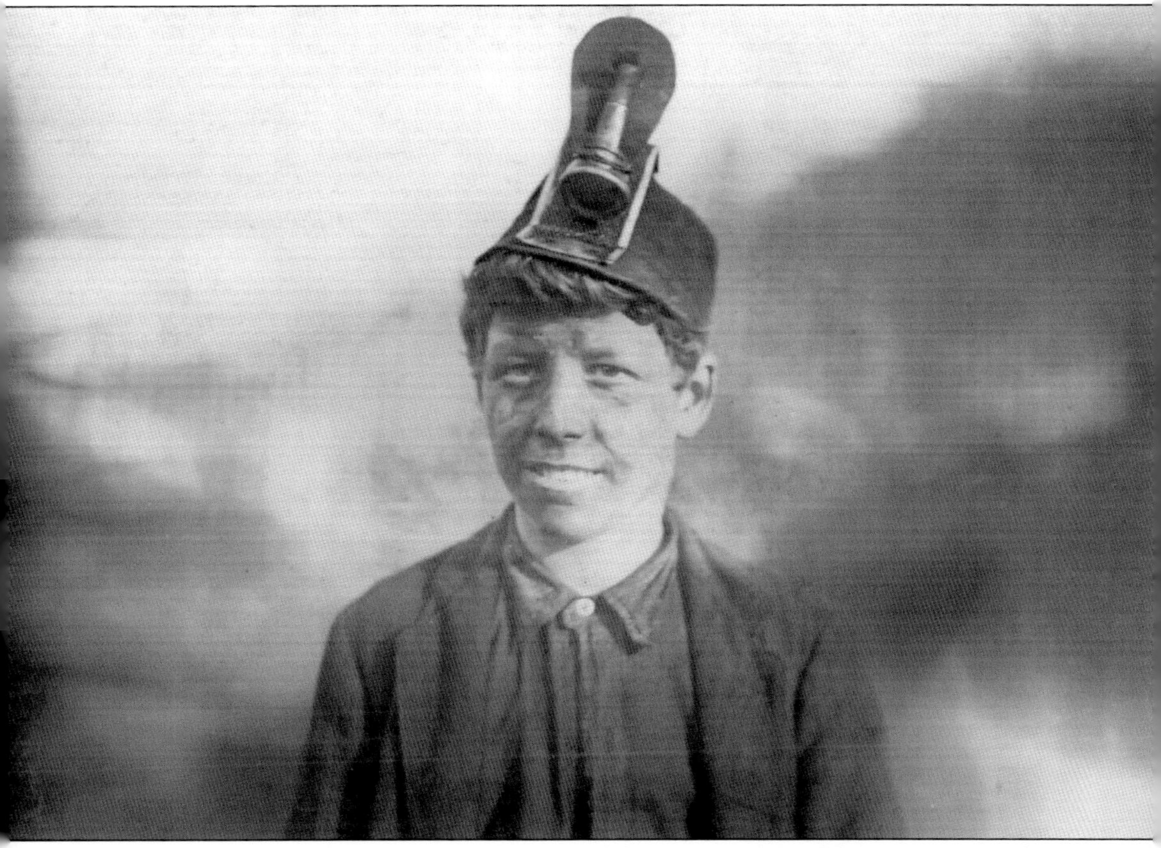

This young miner, named Frank, represents many of the young miners of the era. Lewis Wickes Hines took his photograph in 1907 and recorded that Frank helped his father shovel and load coal. He was hospitalized for one year when a coal car crushed his leg. (LOC, NCLC.)

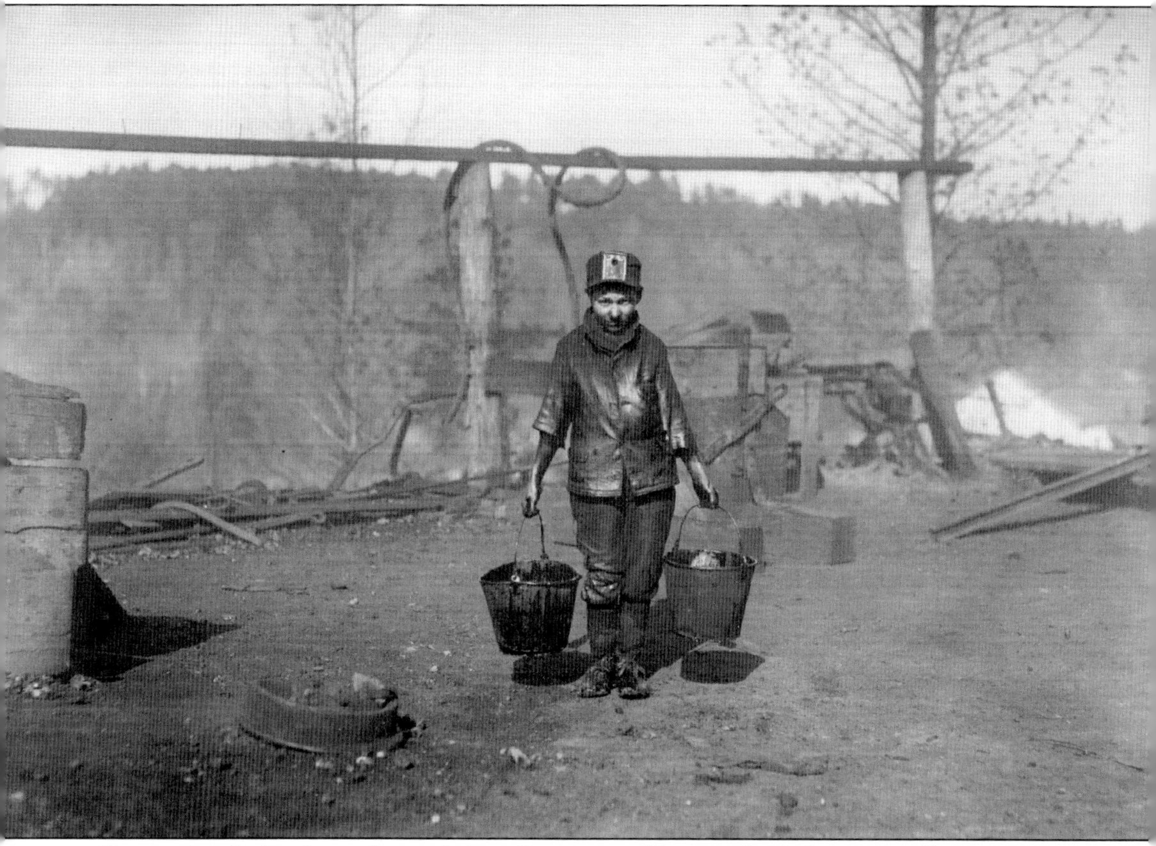

Diamond in the Coalpit author and Pratt City native Clement A. Taylor wrote of his experiences working in the mines at the tender age of 12. He left school to work as a greaser, earning a paltry $3 per month. Taylor later served in the First World War and moved to New York City. (LOC, NCLC.)

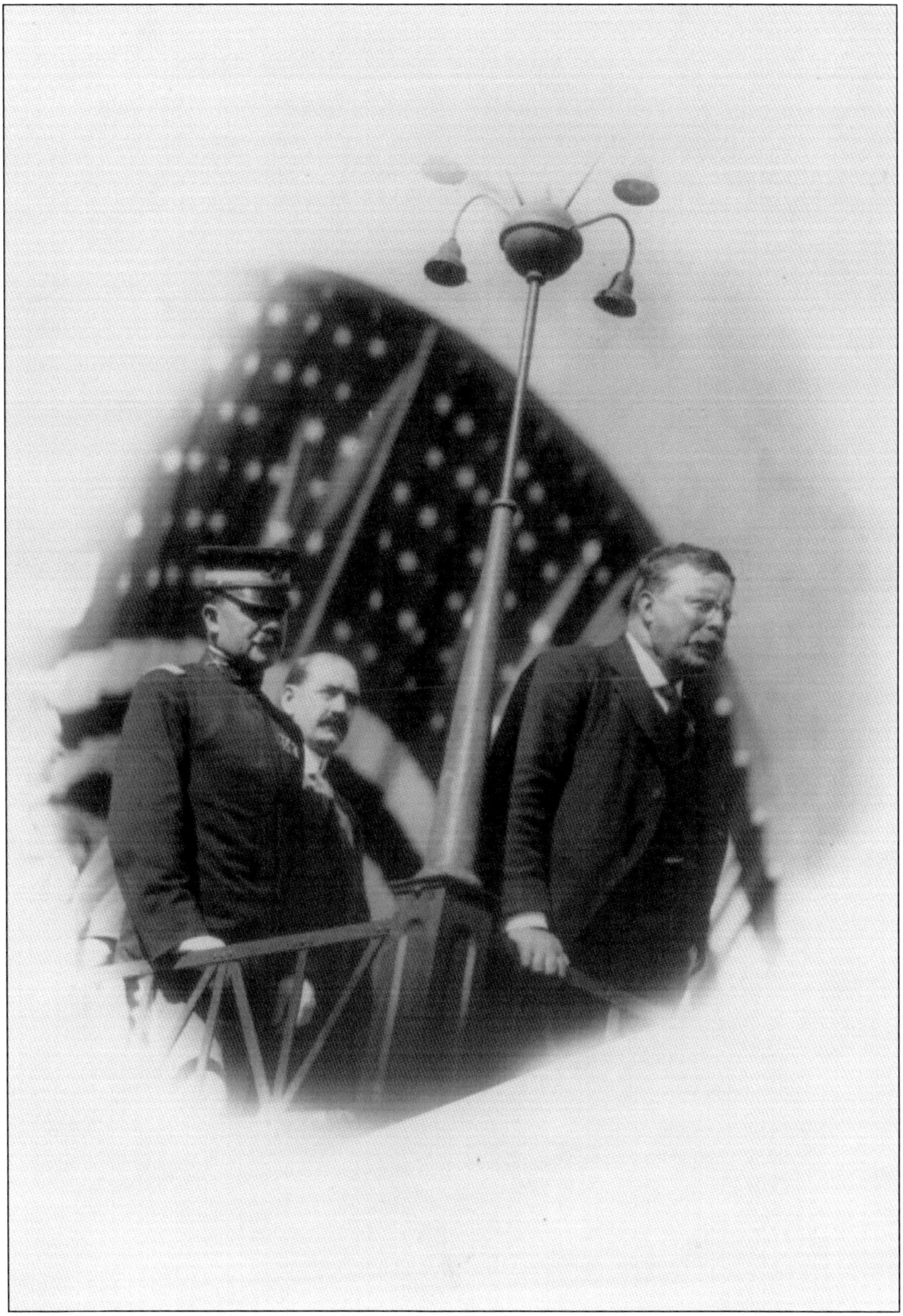

Sickly as a child, Franklin Roosevelt felt a special empathy for children. While the child labor laws did not change during his presidency, the progressive Roosevelt called for a New Nationalism program that would end child labor. (LOC, NCLC.)

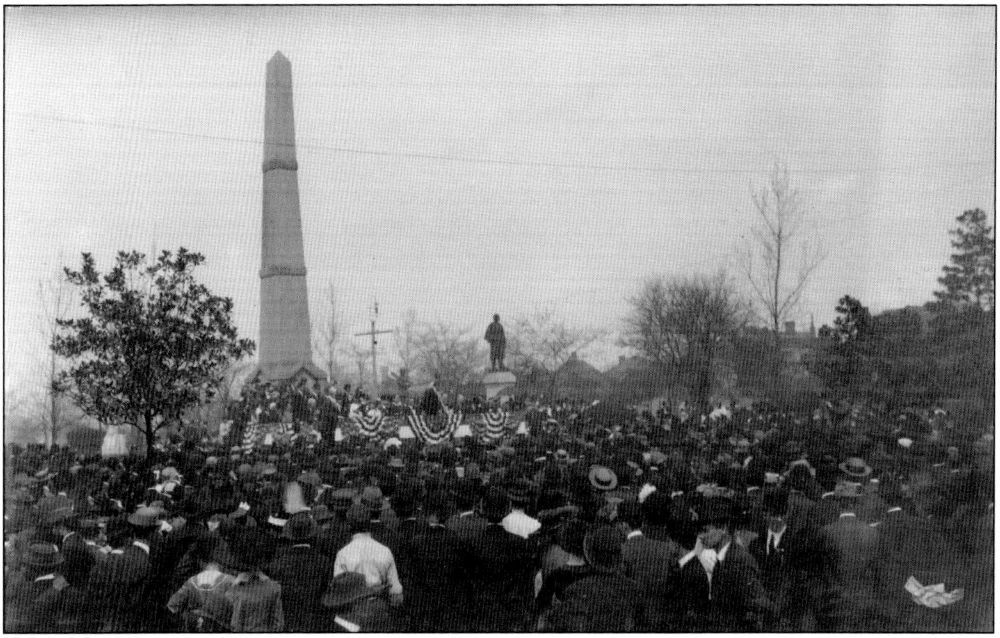

In *The Conservation of Womanhood and Childhood*, Roosevelt wrote, "It is a scandal to the American Nation that our national government has proved so supine in doing what it can to promote the welfare of the children of the country." He speaks here in Birmingham. (LOC, NCLC.)

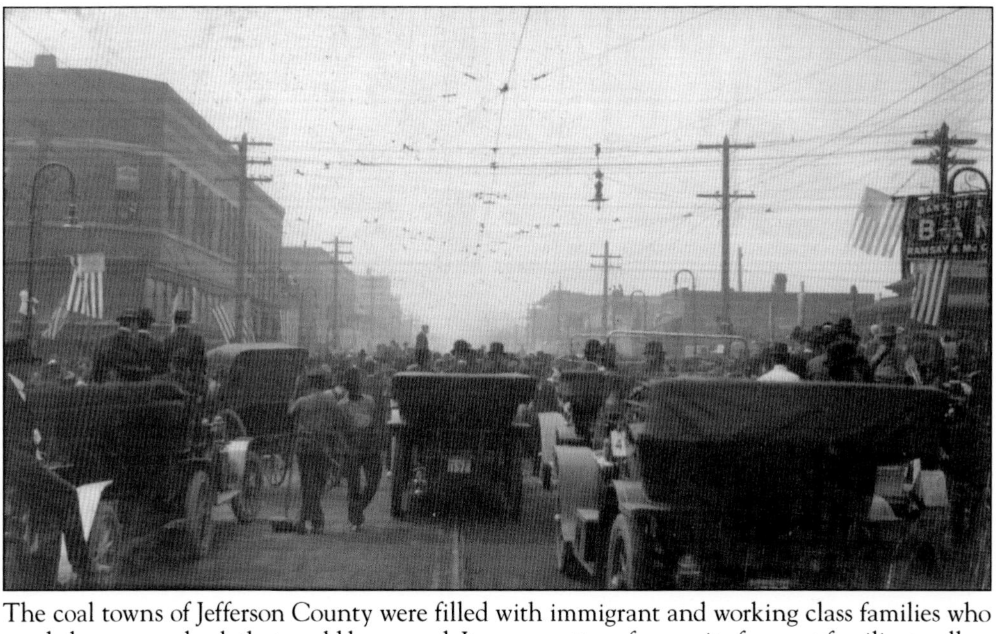

The coal towns of Jefferson County were filled with immigrant and working class families who needed every paycheck that could be earned. It was a matter of necessity for most families to allow their young sons to work in the mines beside their fathers. Despite the large crowds that turned out to hear Roosevelt speak, little came of child labor legislation in Alabama. (LOC, NCLC.)

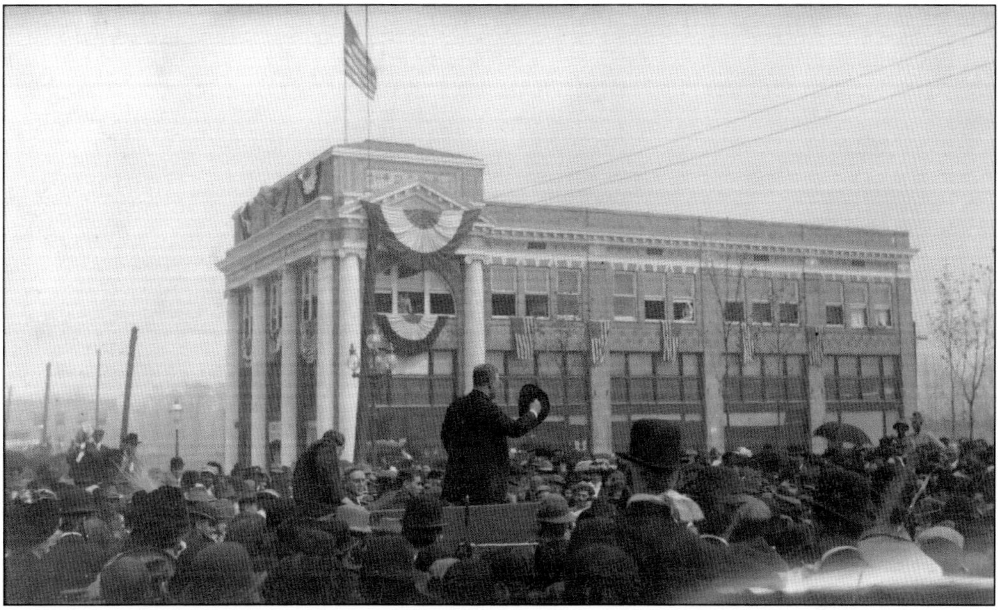

Speaking in Fairfield, then known as Corey, Roosevelt assailed the numbers of children allowed to work long hours for little pay. In 1909, Alabama passed child labor reform that limited the age of child workers to 12 years and the maximum number of hours worked to 60. (LOC, NCLC.)

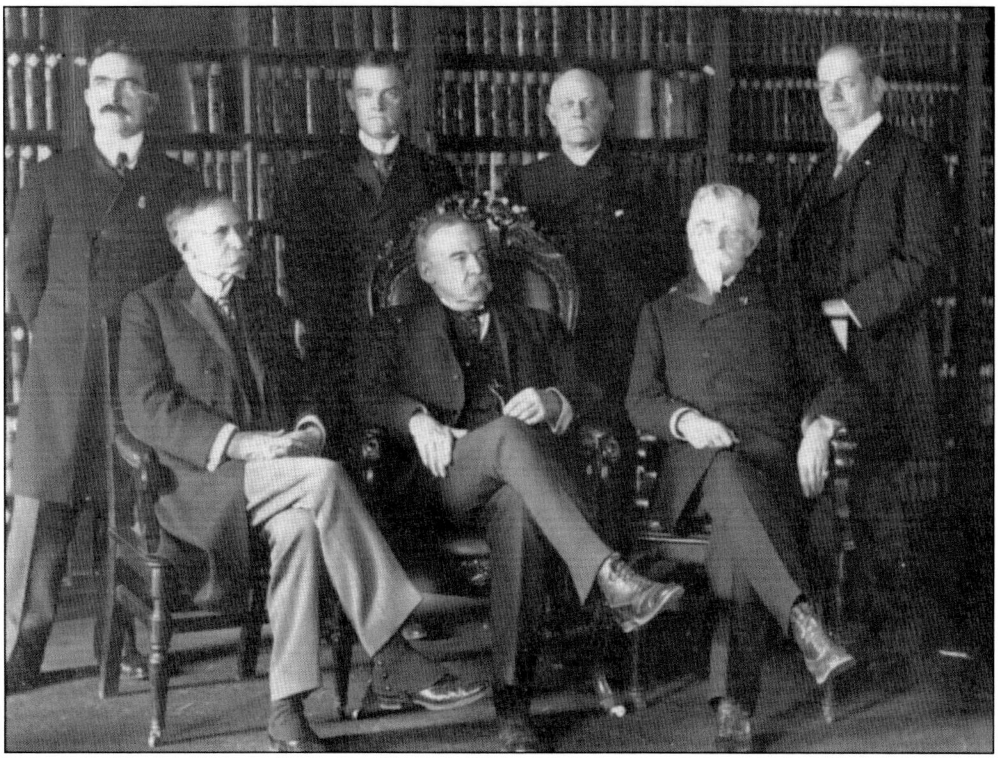

Roosevelt was not only involved in the issue of child labor in the nation's coal mines, he also intervened in 1902. The coal shortage of 1902, caused by a major strike, tempted Roosevelt to send the Army to mine coal. Here, Roosevelt assembles a team to arbitrate the strike. (LOC.)

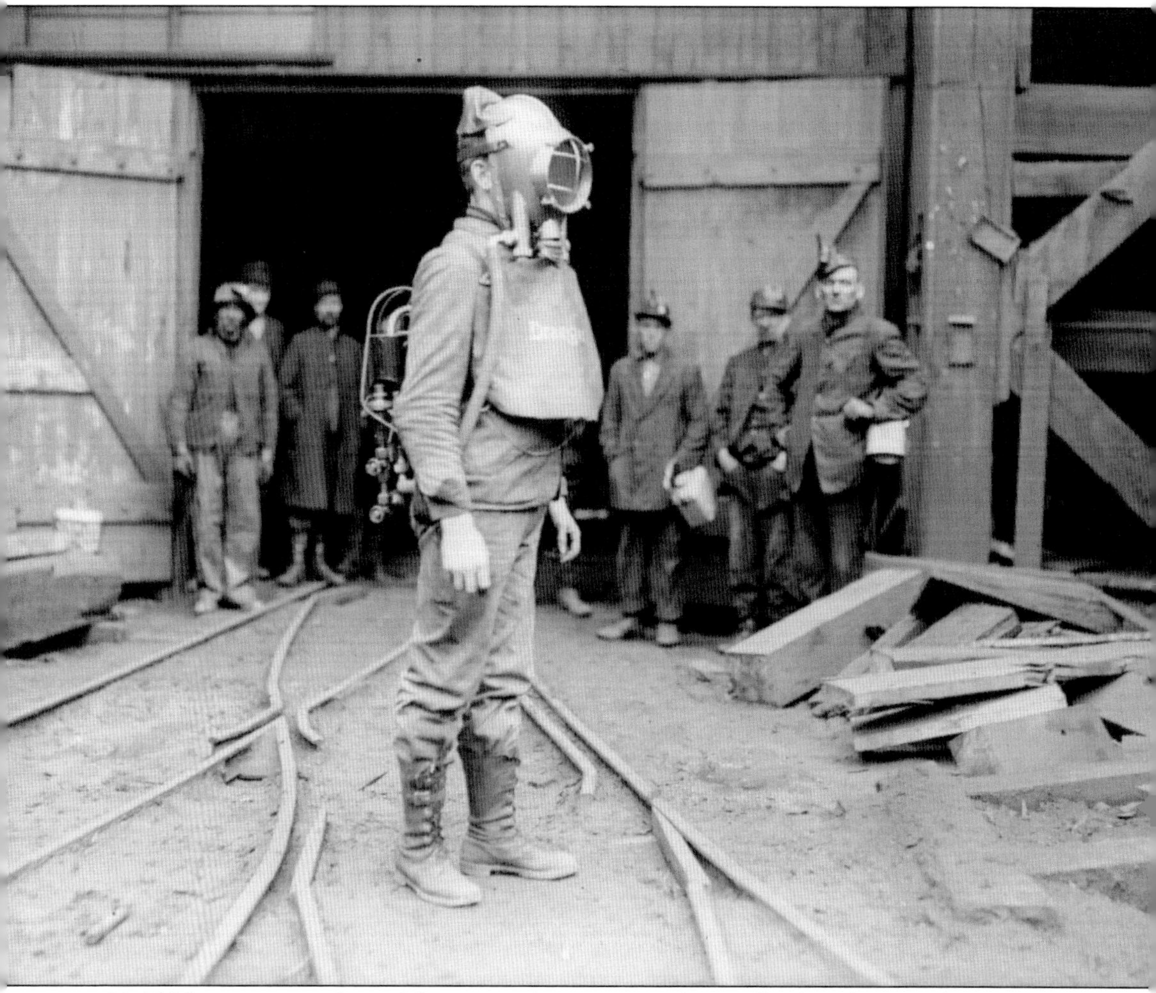

Child labor was not only used for mining, but rescue work as well. This photograph horrified the nation. (LOC, NCLC.)

America's entry into the First World War necessitated mobilization of the country's economy to a wartime one. Pres. Woodrow Wilson appointed Stanford academic Herbert Hoover to head the government's Fuel Administration. (LOC.)

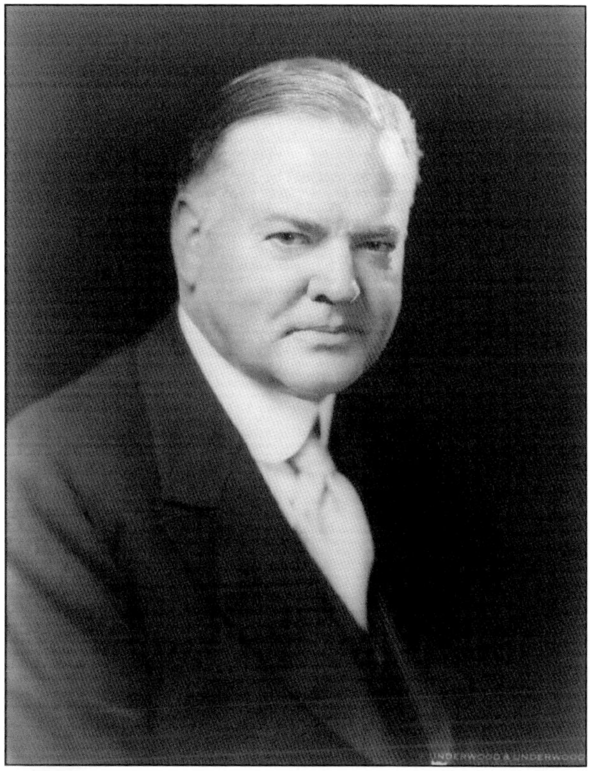

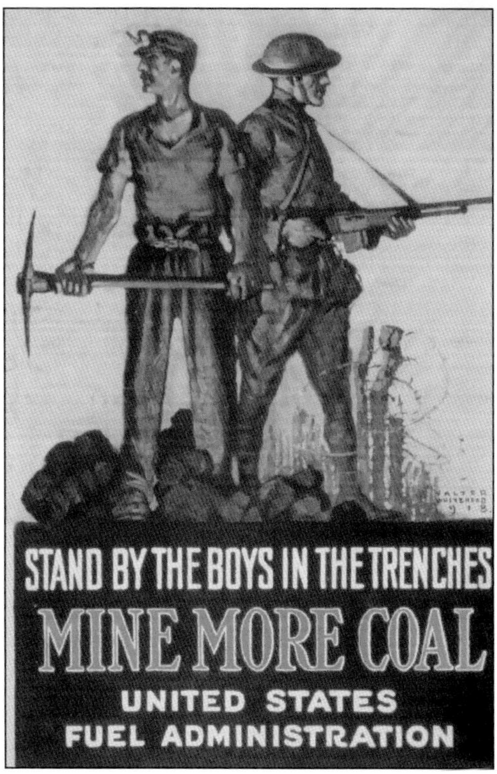

The Fuel Administration designed a series of posters that encouraged America's mine workers to do their part to fuel the United States war machine. Patriotism was aligned with production, and miners across the country stepped up. Unfortunately, as the economy shifted again after the war, coal operators wished to cut back on rates, leading to a massive strike in 1920. (LOC.)

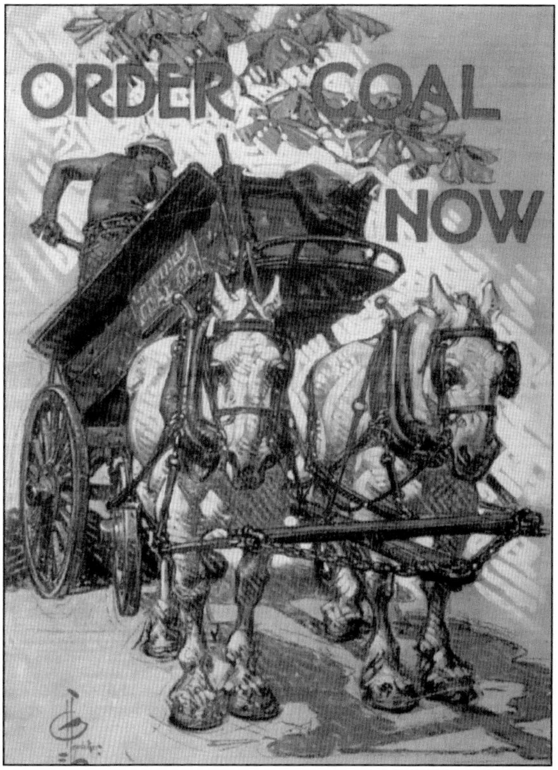

During the middle of the Second World War, the United States government urged people to buy more coal, which implied that miners should mine more coal. Interestingly, 1943 would be a year of strikes and disasters at Jefferson County mines. (LOC.)

Miner-operator relations were often strained to the point of violence. In July 1908, three military companies, 10,000 rounds of ammunition, and a Gatling gun were sent to Blossburg and Republic in order to curb labor-management tensions. (ADAH.)

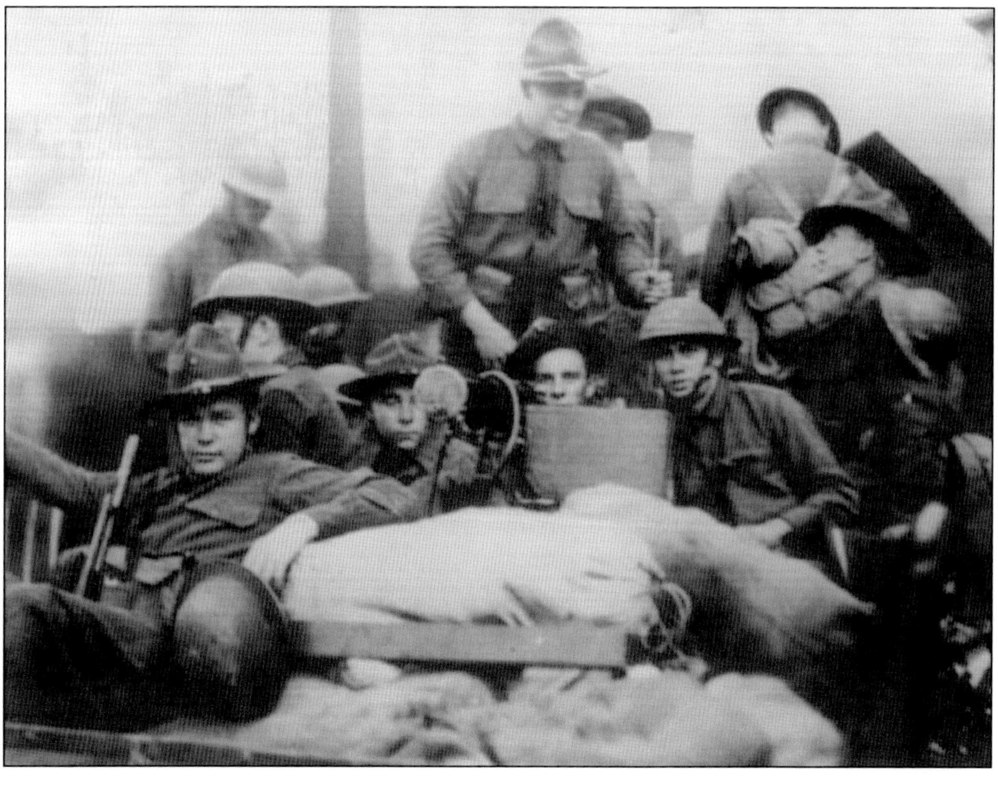

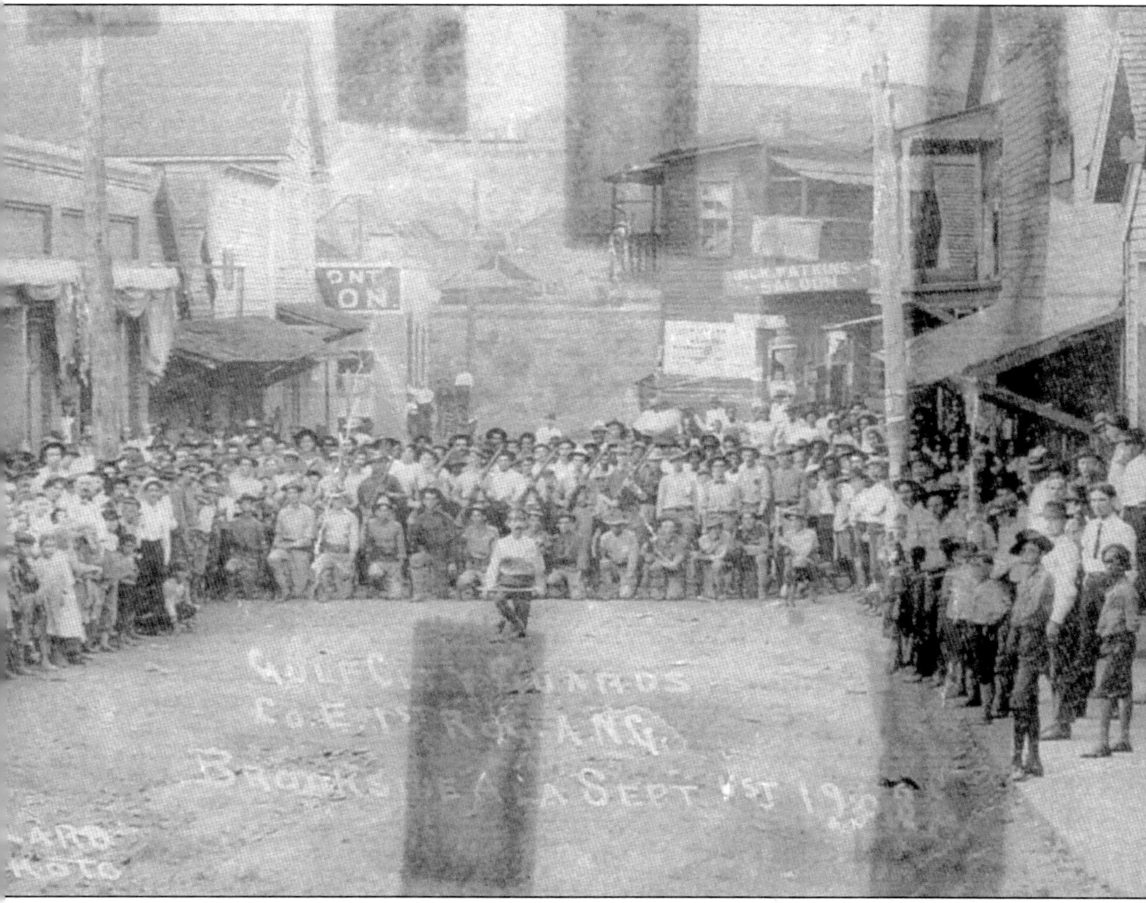

The 1908 strike was due to coal operators attempt to put into place a 17 percent rate reduction. When miners struck, owners brought in "black legs" to restart production. Governor Comer sent in the Gulf States Guards to Brookside to keep the peace. (BHM.)

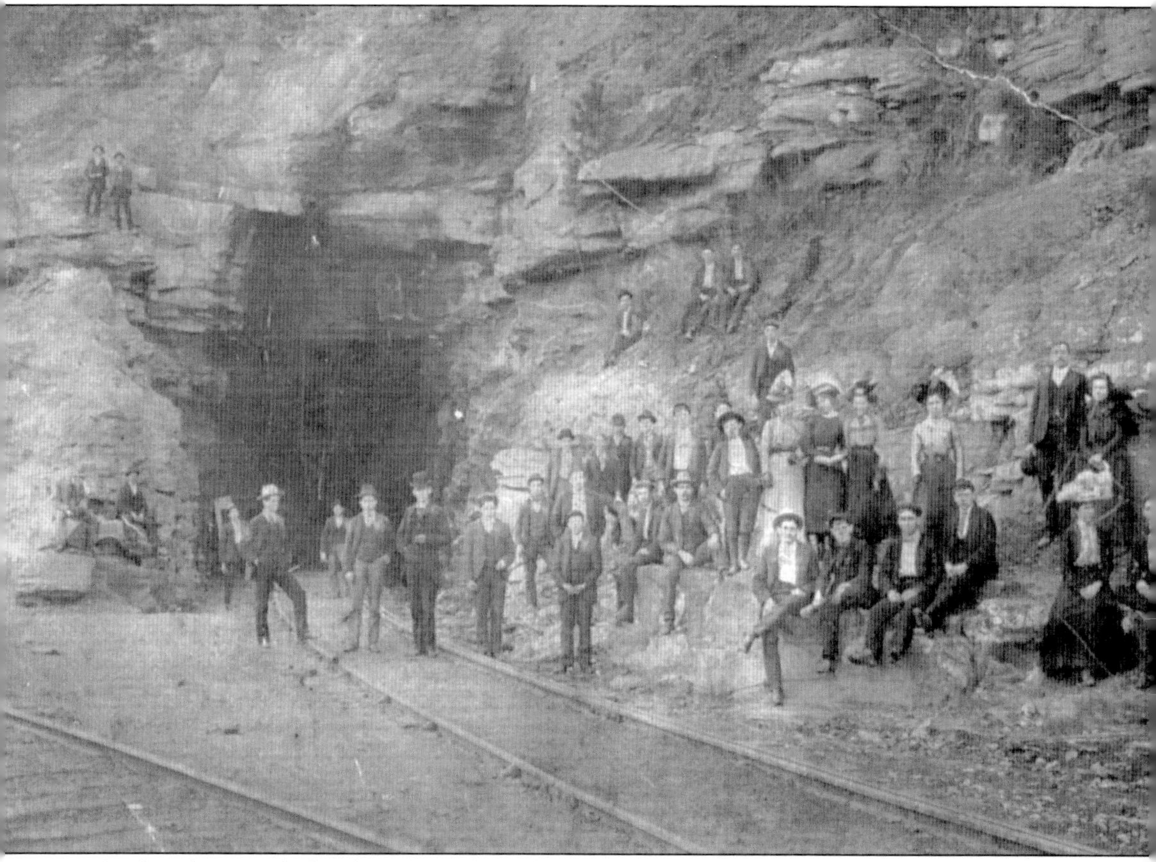

In this photograph, local residents dressed in their Sunday best enjoy an afternoon at Jefferson Tunnel near Cardiff. In 1908, times were not so peaceful when miners took to the surrounding hills and waited for a train filled with strikebreakers to exit the tunnel. As it emerged from the tunnel, miners armed with rifles opened fire. (BHM.)

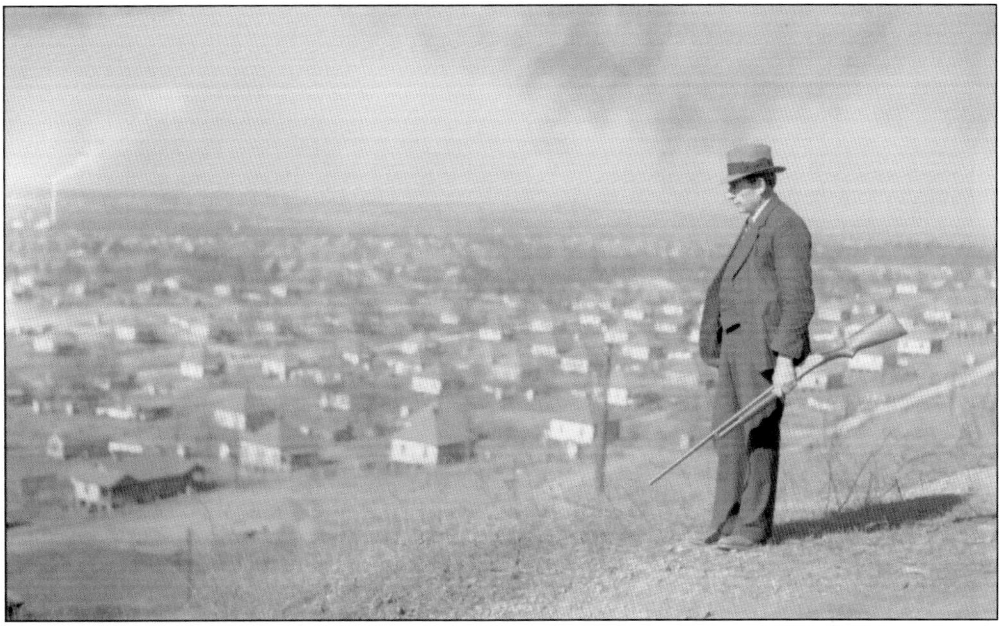

This company guard looks menacingly over a town during strike times. (LOC, FSA.)

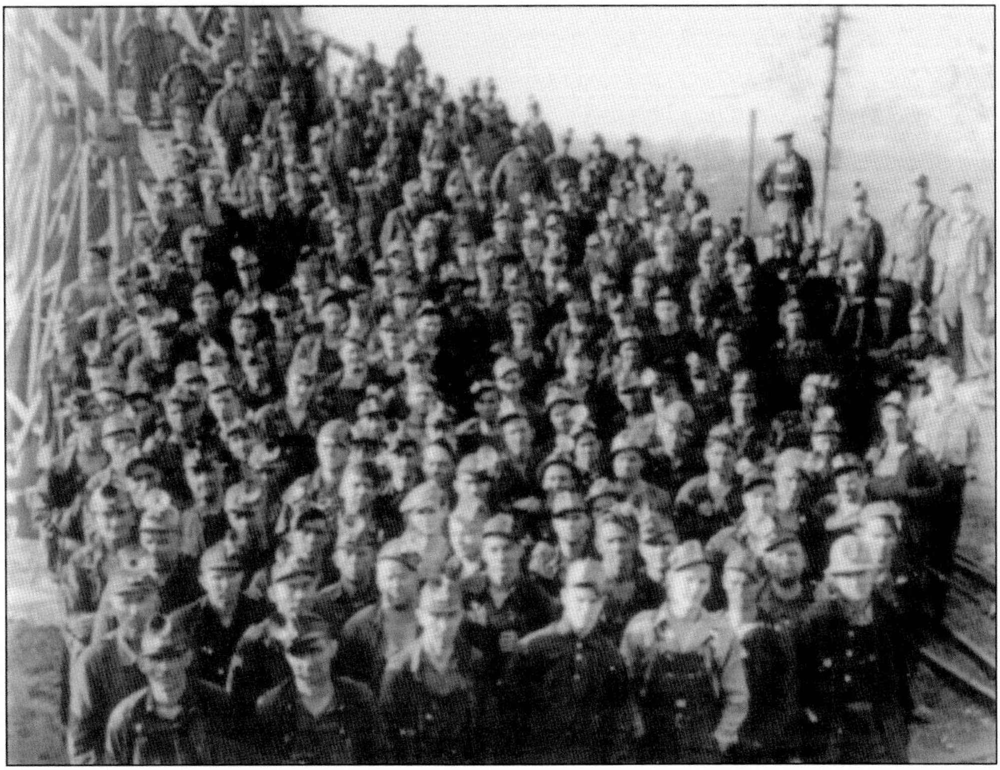

For years, many believed racial strife among postbellum blacks and whites to be the decisive factor in why miners had difficulty unionizing easily. Racial strife was more a creation of coal operators than reality among miners. However, racial tensions did spill over. In Cardiff in 1898, two black miners were killed by whites. (ADAH.)

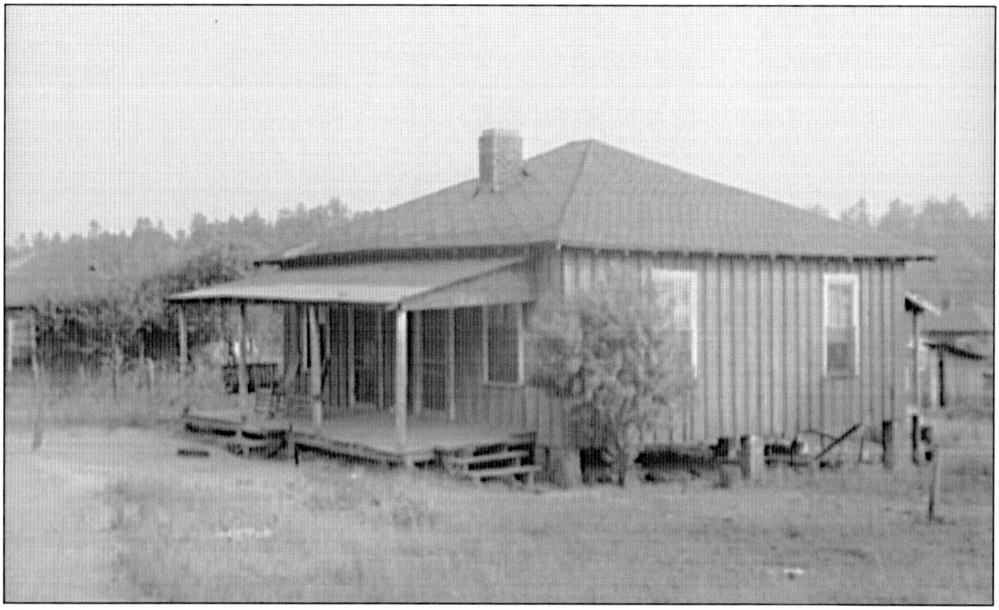

This photograph shows a quiet Bradford homestead. Bradford would become the scene of intense labor strife in a few short years. (Horace Defore.)

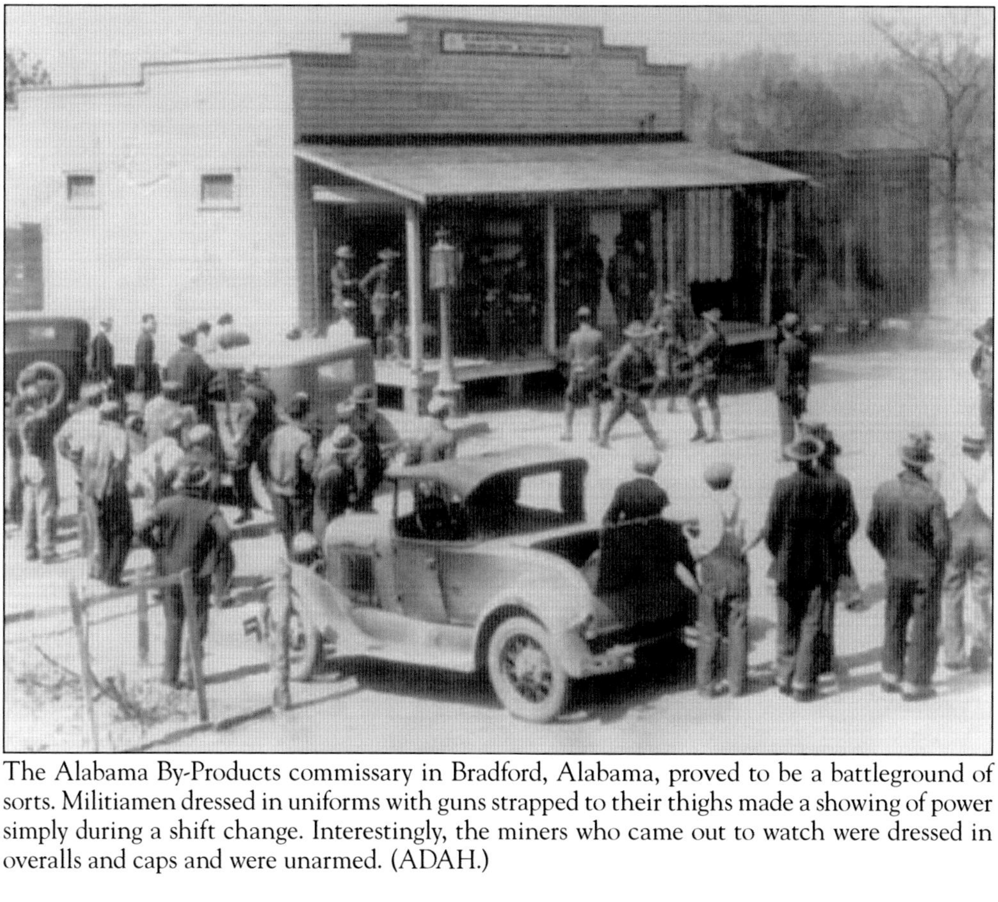

The Alabama By-Products commissary in Bradford, Alabama, proved to be a battleground of sorts. Militiamen dressed in uniforms with guns strapped to their thighs made a showing of power simply during a shift change. Interestingly, the miners who came out to watch were dressed in overalls and caps and were unarmed. (ADAH.)

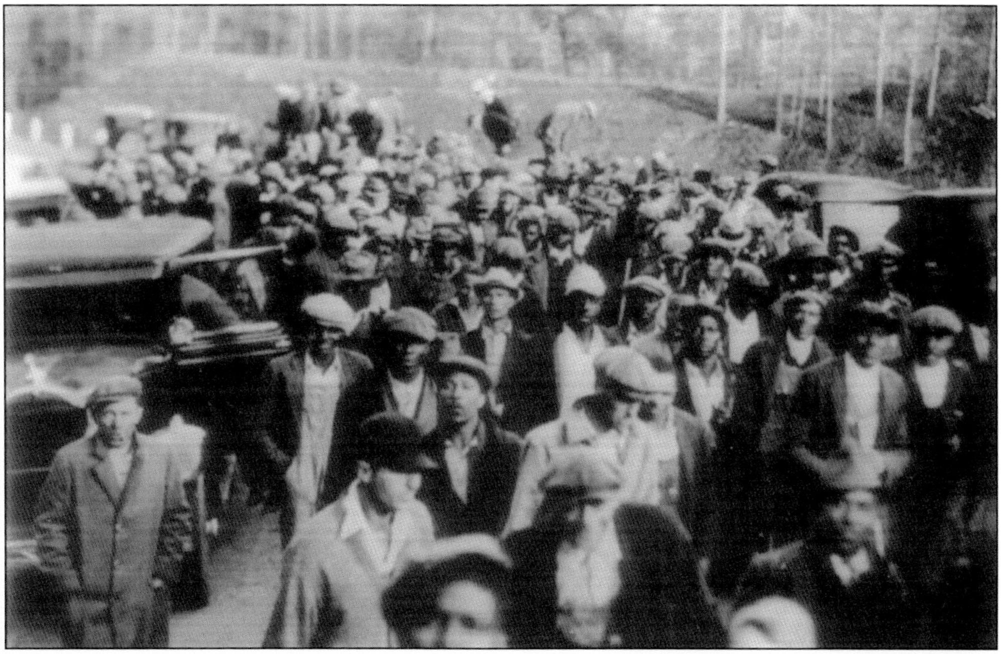

In 1934, twelve hundred miners struck at Edgewater and more demonstrated at the Red Diamond mines. The governor called in 75 members of the state militia, who had machine guns directed at the crowd. Newspapers reported that one picketer died at Red Diamond, which had continued to operate in spite of the strike. (ADAH.)

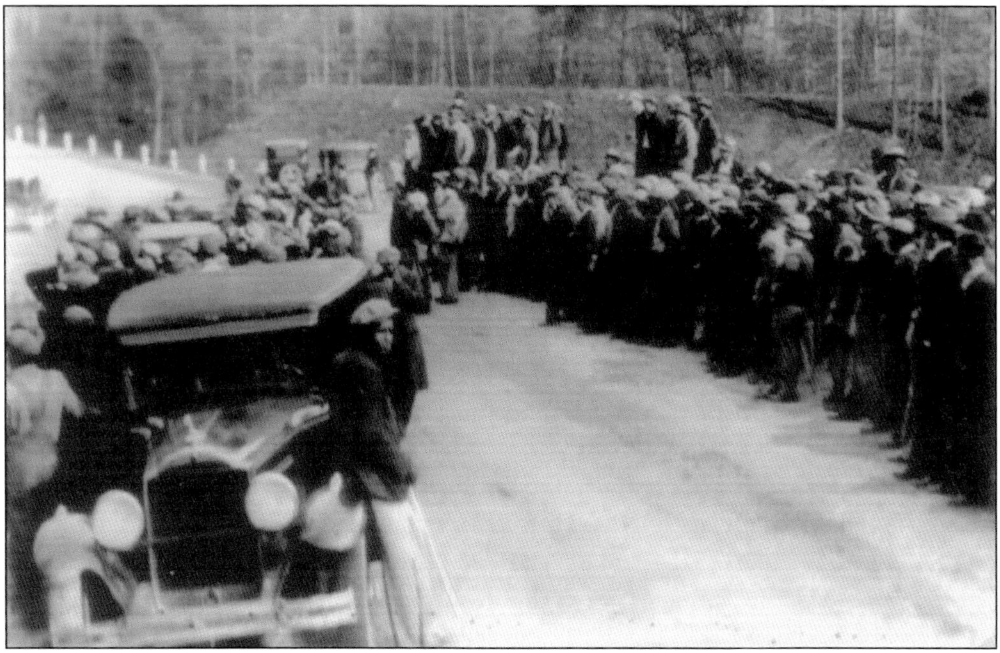

Striking miners at Edgewater in 1934 were alleged to have attempted to destroy water mains at both TCI's Ensley and Fairfield steel plants. One hundred guards remained in Edgewater to quell the impending storm. In addition, two men were arrested for passing out communist flyers to the strikers. (ADAH.)

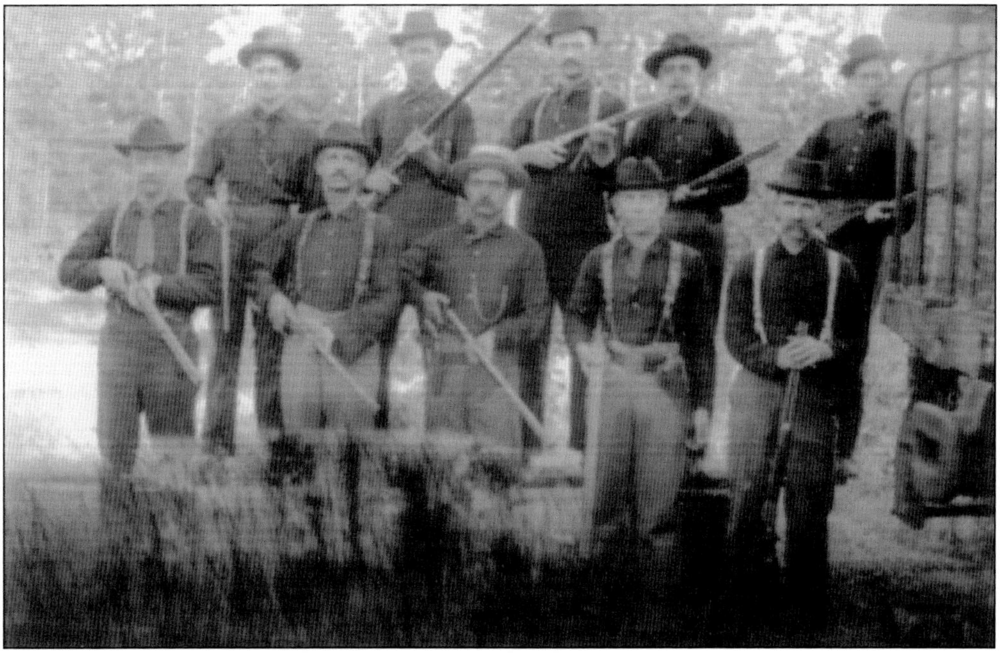

One of the last great coal strikes broke out in 1920. The strike helped end the great heyday of coal mining in Jefferson County. Many left the area looking for jobs. John Sokira and his son temporarily moved to Kentucky to find work. Others, like the Mulkin family from Cardiff, made the permanent move to Illinois. (ADAH.)

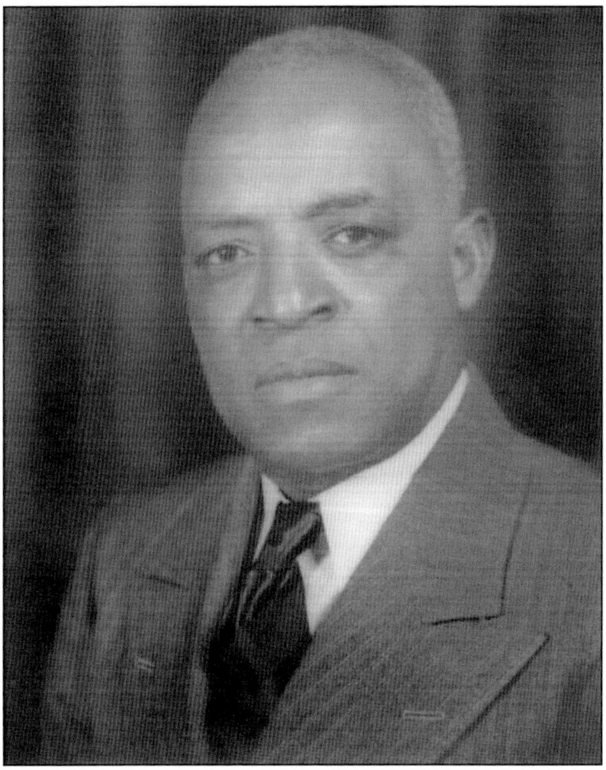

Though the United Mine Workers of America (UMW) had organized in Alabama in 1898, societal, governmental, and economic factors worked against any progress by the union. The perils of the Great Depression and the energies of Montevallo's Walter W. Jones helped reorganize the UMW in Alabama into a powerful entity during the middle part of the century. (ADAH.)

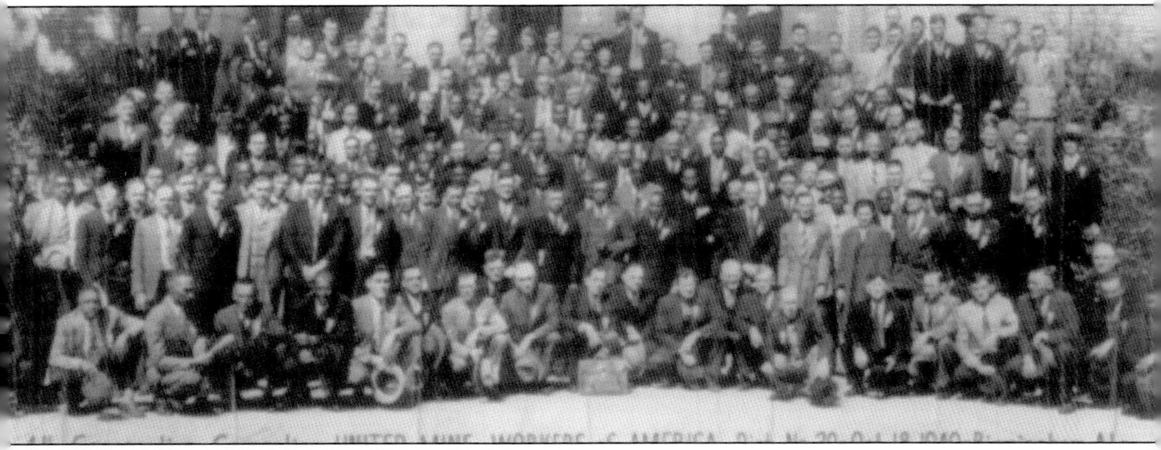

The challenges of Alabama coal miner unionization have long fascinated historians. In the early years of unionizing attempts, miners wondered which organization would best aid them. The Knights of Labor appealed to miners of both races, but it was the United Mine Workers of America that organized the miners and pushed for better pay, safer conditions, and an end to convict leasing. (ADAH.)

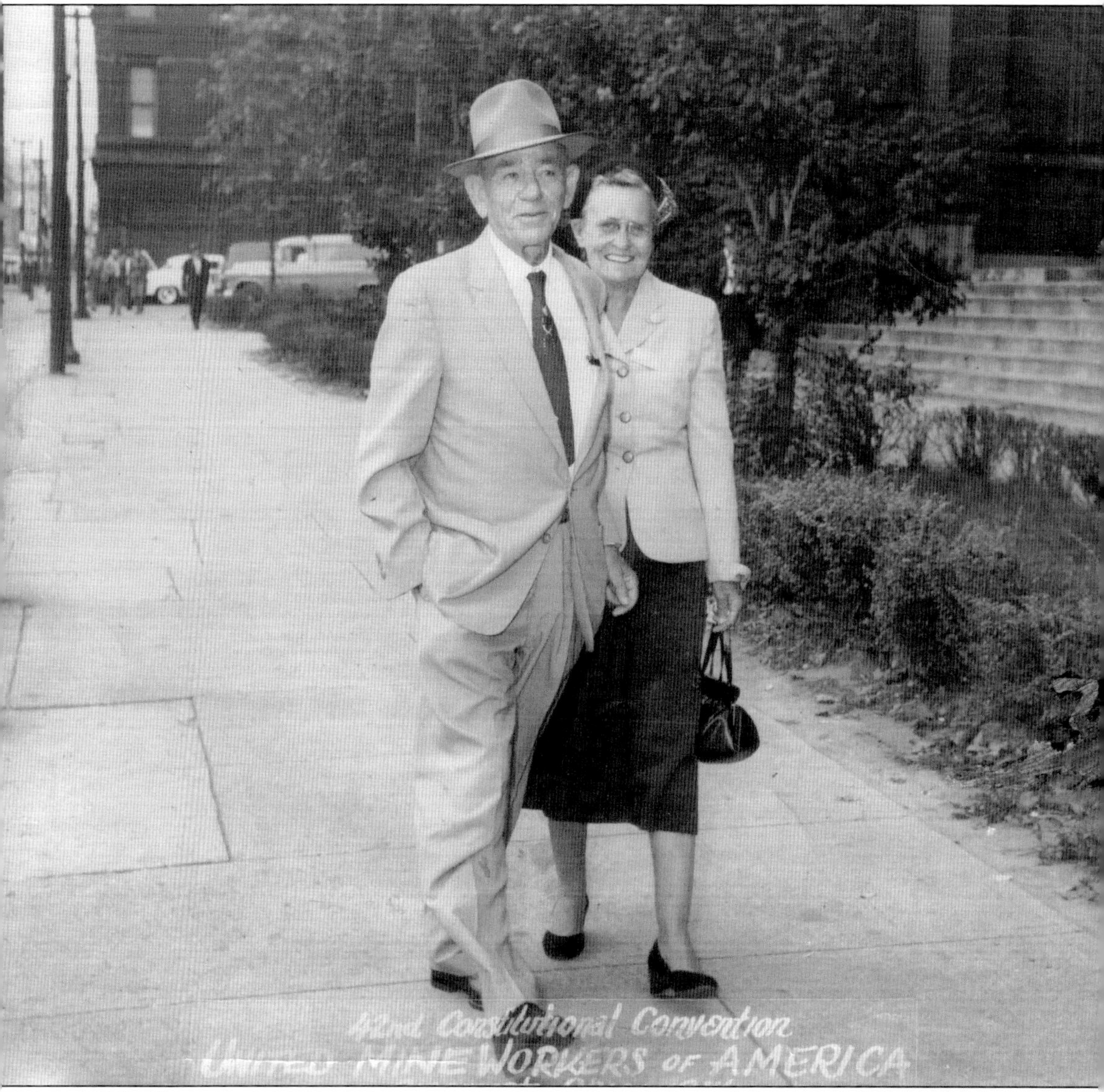

Ed Billits, whose parents were immigrants into Brookside, served District 20 for many years as a District 20 representative to the national UMW meetings. (BHM, Albert Billits Collection.)

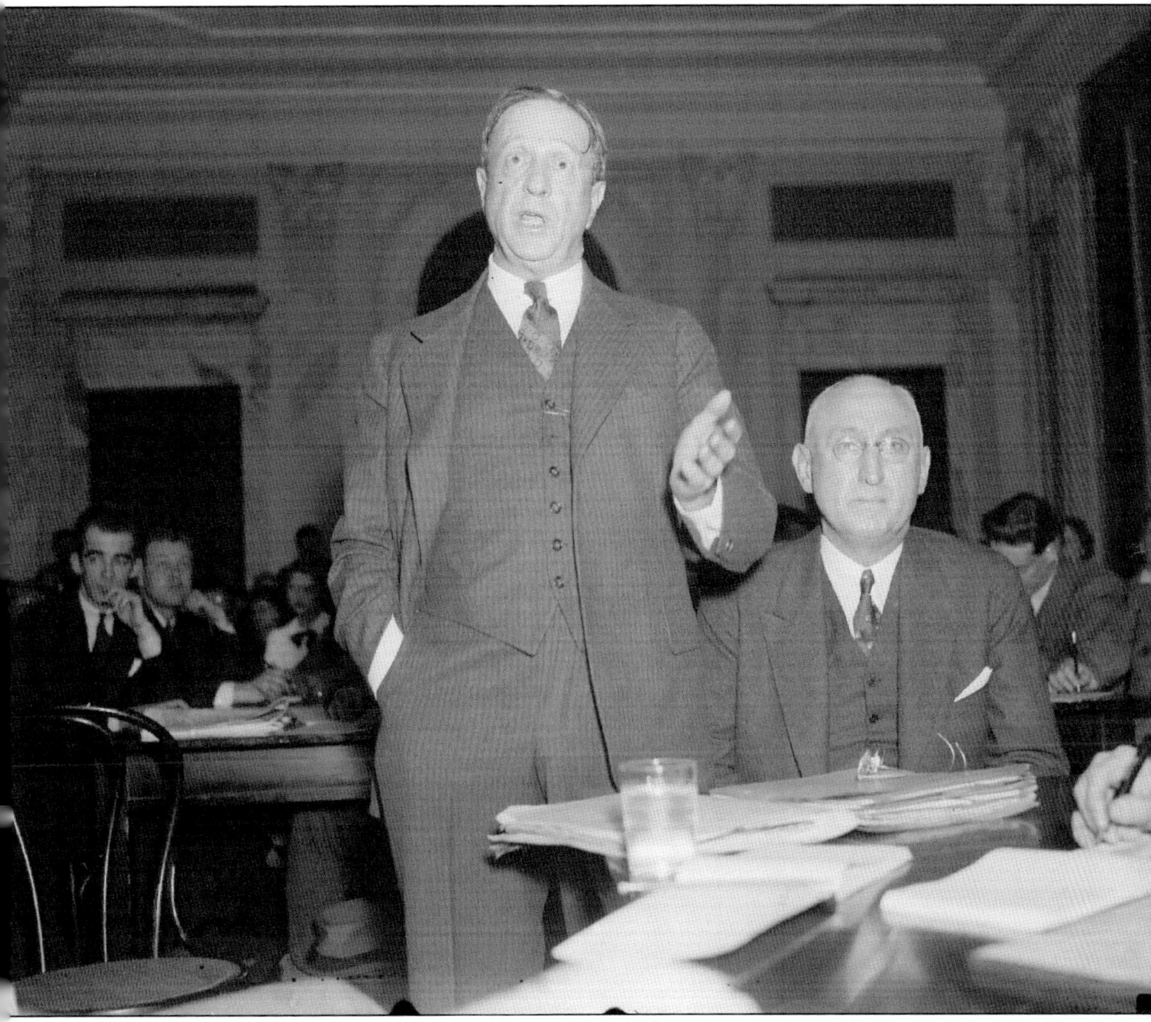

This 1937 photograph shows TCI counsel testifying before Congress on allegations that coal company guards flogged miners during a strike. Attorney Borden Burr contended that TCI was forced to hire guards to protect lives and property after the failure of local law enforcement to do so. (LOC.)

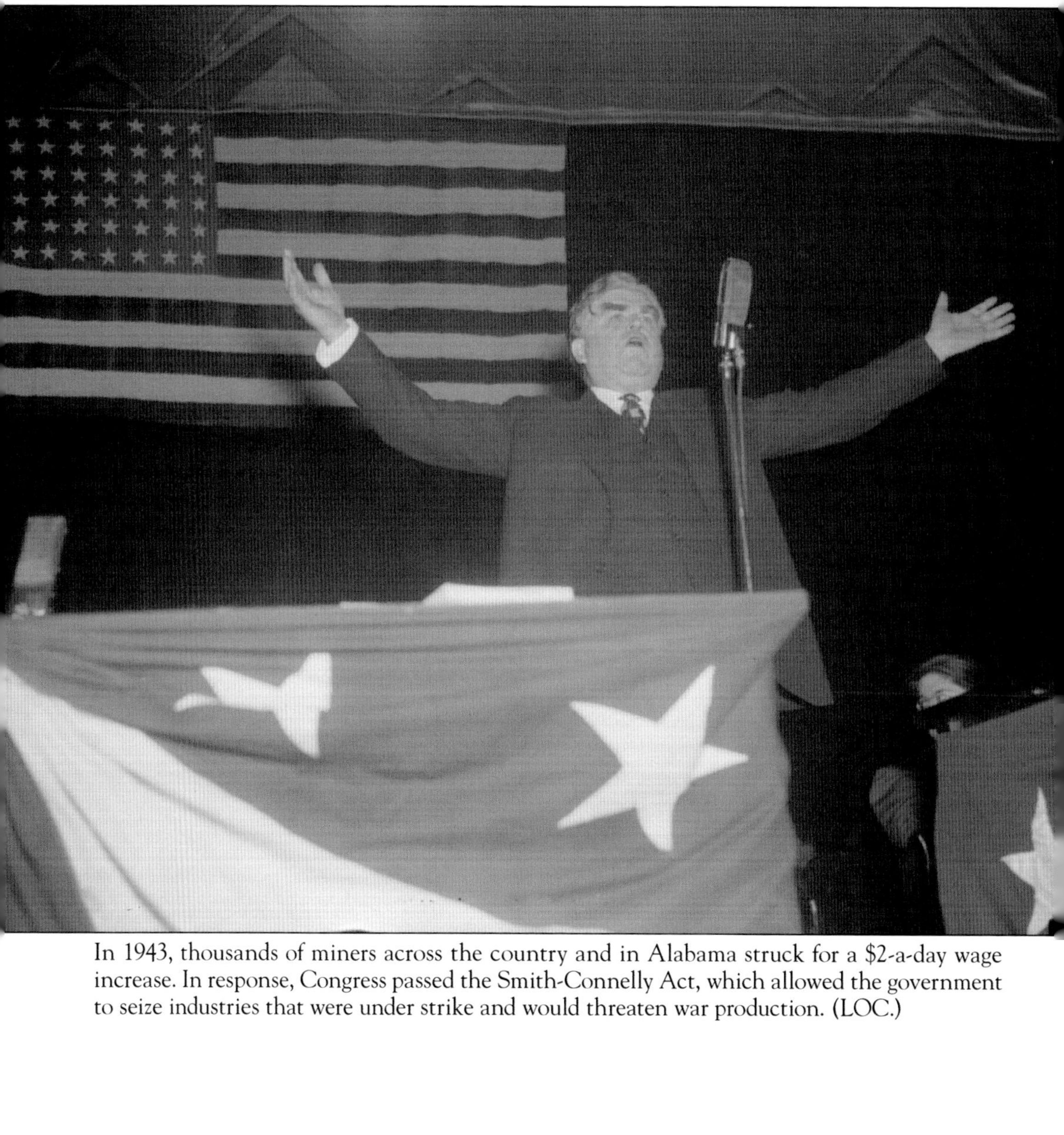

In 1943, thousands of miners across the country and in Alabama struck for a $2-a-day wage increase. In response, Congress passed the Smith-Connelly Act, which allowed the government to seize industries that were under strike and would threaten war production. (LOC.)

Three

How They Lived

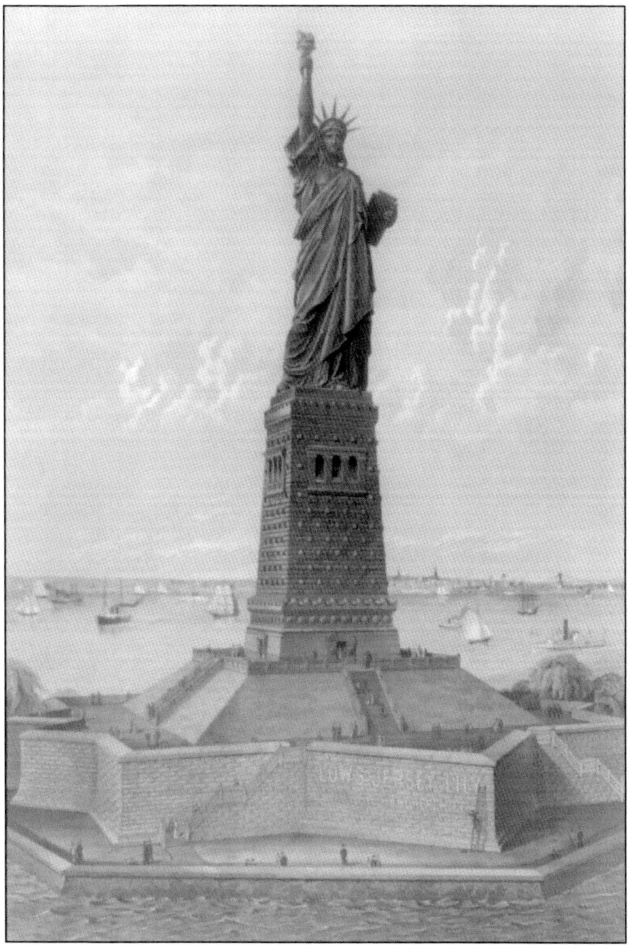

One of the great magnetic factors to the United States was the attraction of freedom and employment. Many of Jefferson County's coal miners entered America through Ellis Island, New York. Some hoped to find permanent employment, while others wanted to earn enough money to return home to buy a farm. (LOC.)

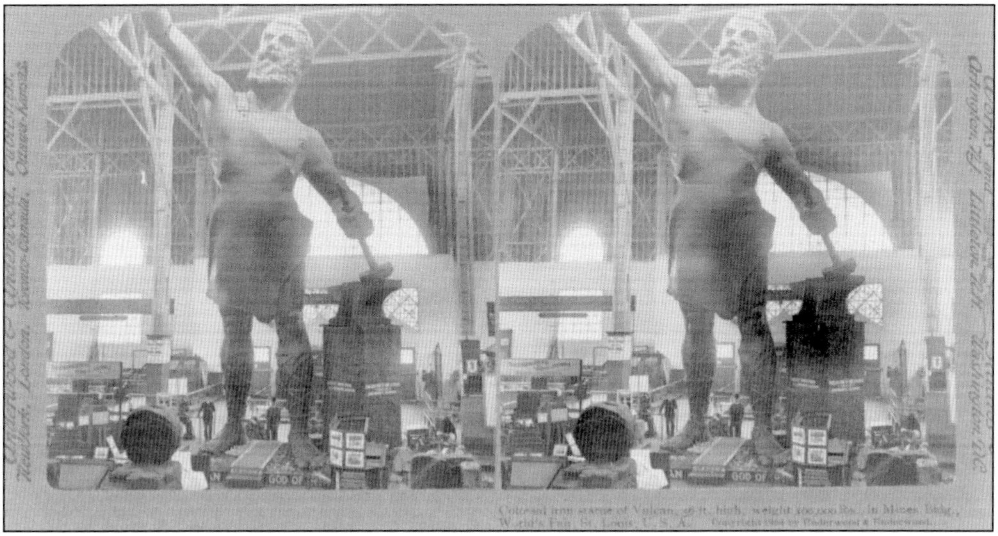

Once immigrants made the trip from New York or other points of entry, they entered the "Magic City" built on iron, steel, and coal. Many were shocked at the condition of the New South City. Nellie Saia from Sicily complained of mud roads and "nasty shoes," something she had not experienced in her birthplace. (LOC.)

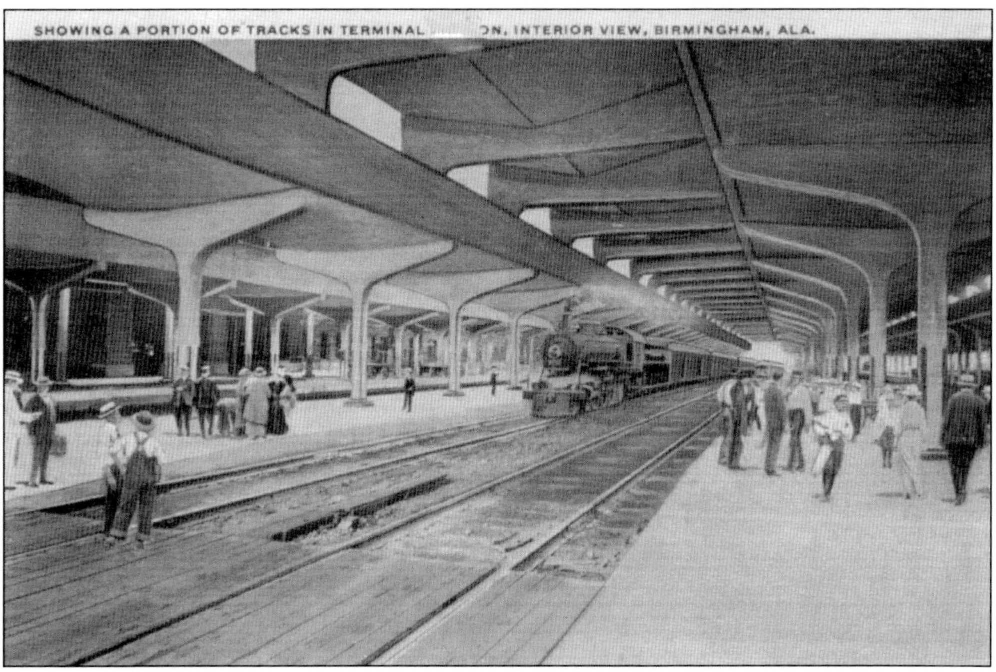

As trainloads of immigrants made their way into Birmingham, they scattered to find their ways into various industries. Many Slovak, Russian, and Scotch-Irish filtered out into the suburbs to places like Brookside, Cardiff, and Pinckney City. (Author's collection.)

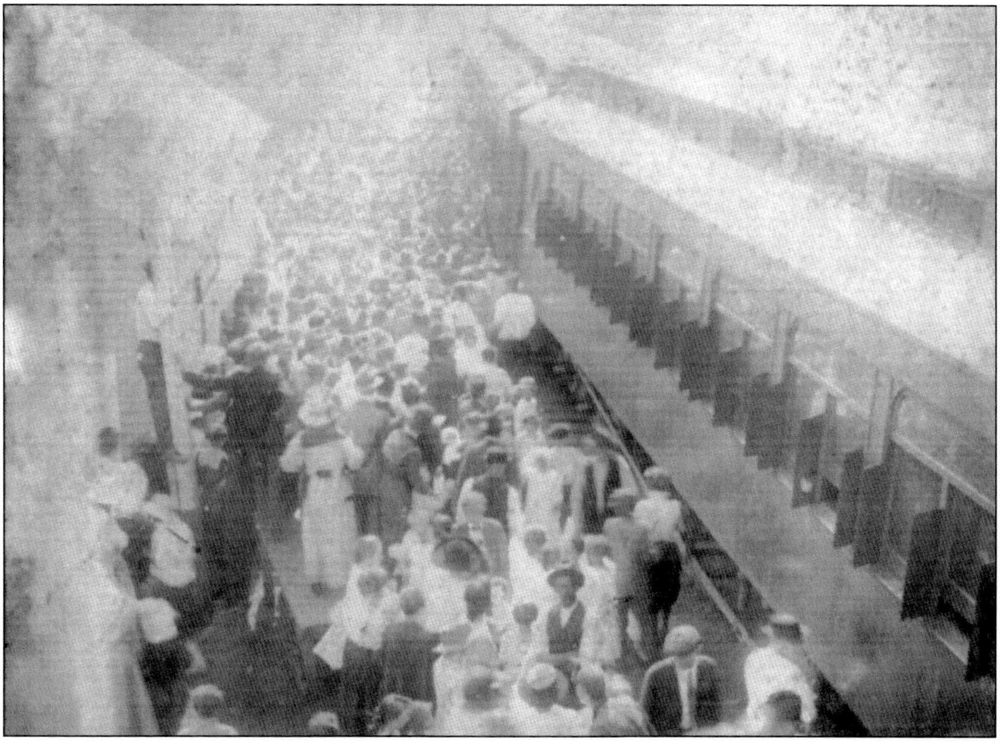

Though this photograph shows a crowded Brookside depot during World War I, Dolphina Lesko from Spis, Austro-Hungary, found herself alone with no one to meet her when she arrived at the same depot years earlier. During the course of the next two years, she married, became a mother, and lost her husband and son to typhoid in 1901. (BHM.)

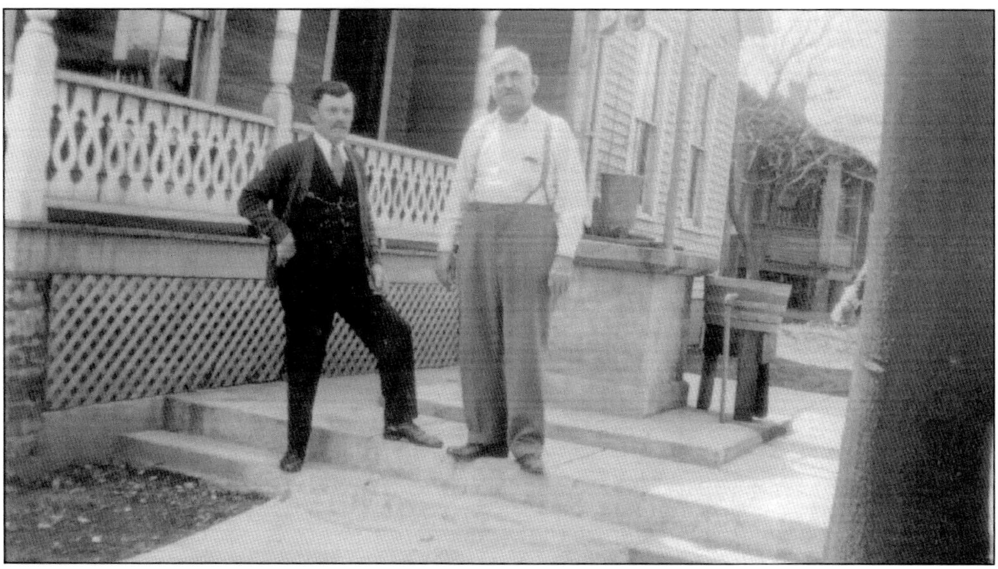

The white-haired gentleman in this photograph is Louis Perunko. Perunko had settled in Brookside when it first opened as a mining camp slightly before the turn of the 20th century. Through his letters home to family, he is credited with encouraging most of Brookside's immigrant families to settle and become miners. (BHM.)

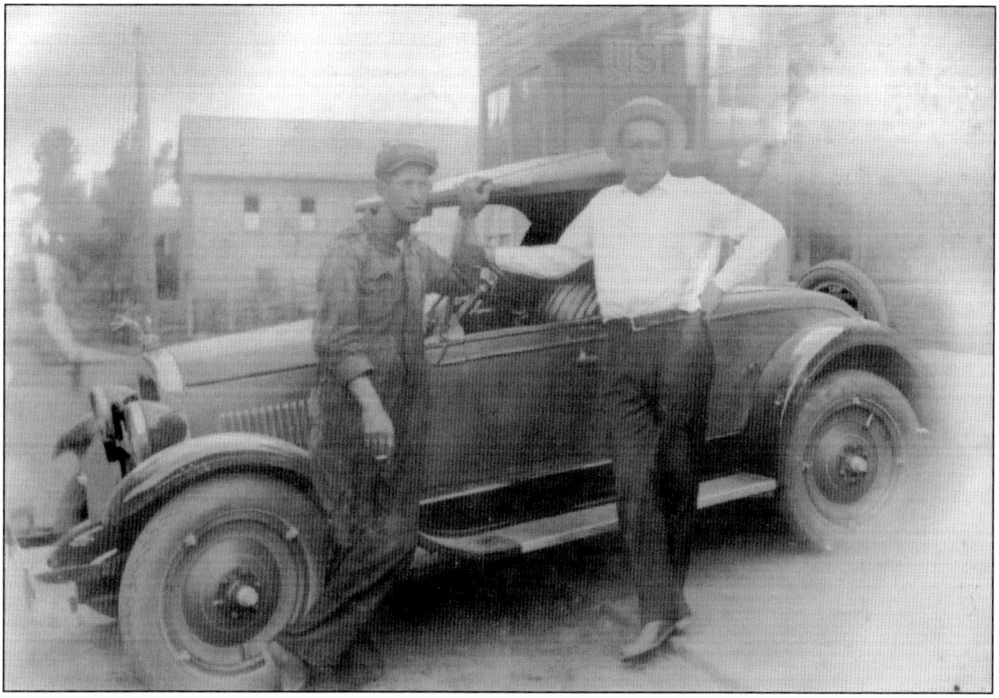

Between 1900 and 1910, immigrants and others filled the Birmingham Industrial District's coalfields. Here are two men enjoying a lazy day in Brookside, one of the busiest of Jefferson County's coal towns. (BHM.)

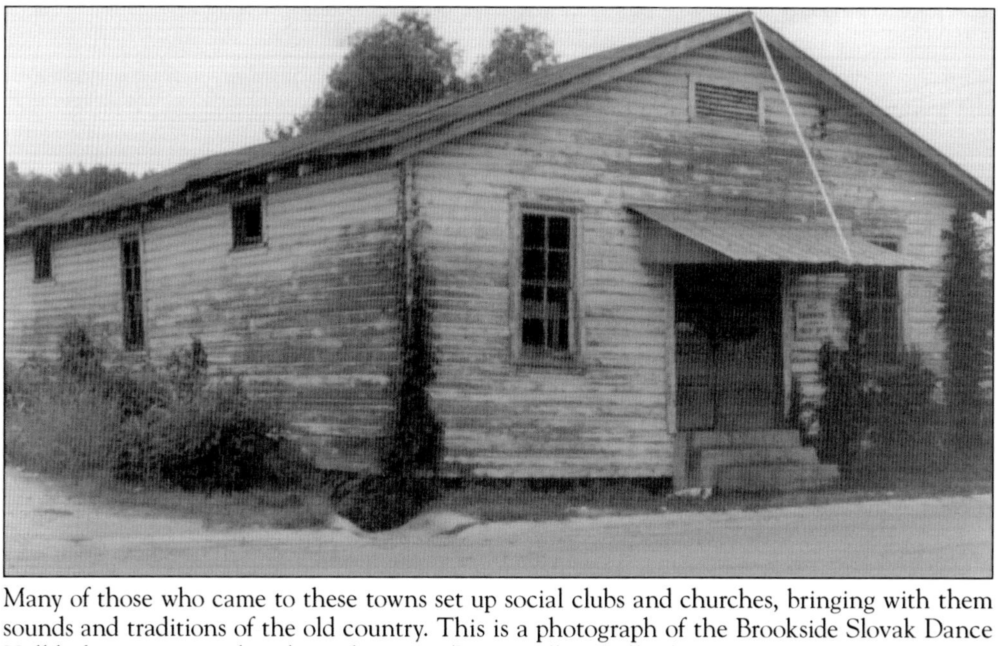

Many of those who came to these towns set up social clubs and churches, bringing with them sounds and traditions of the old country. This is a photograph of the Brookside Slovak Dance Hall before it was razed in the early 1980s. (BHM, Albert Billits.)

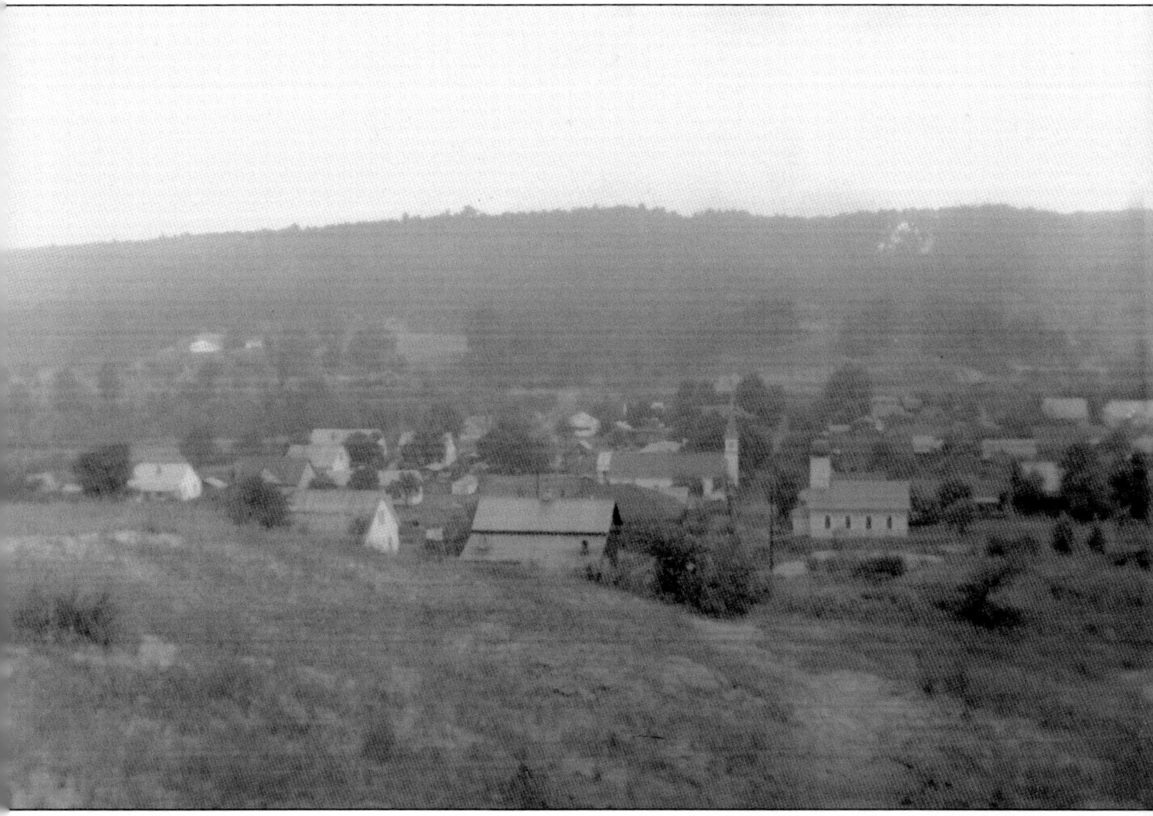

Brookside was divided into two sections that were literally divided by the rail line that hauled coal and coke to other environs. This portrait shows the village of Brookside, which immigrants founded. It was a piece of Europe in the Deep South. (BHM, Albert Billits.)

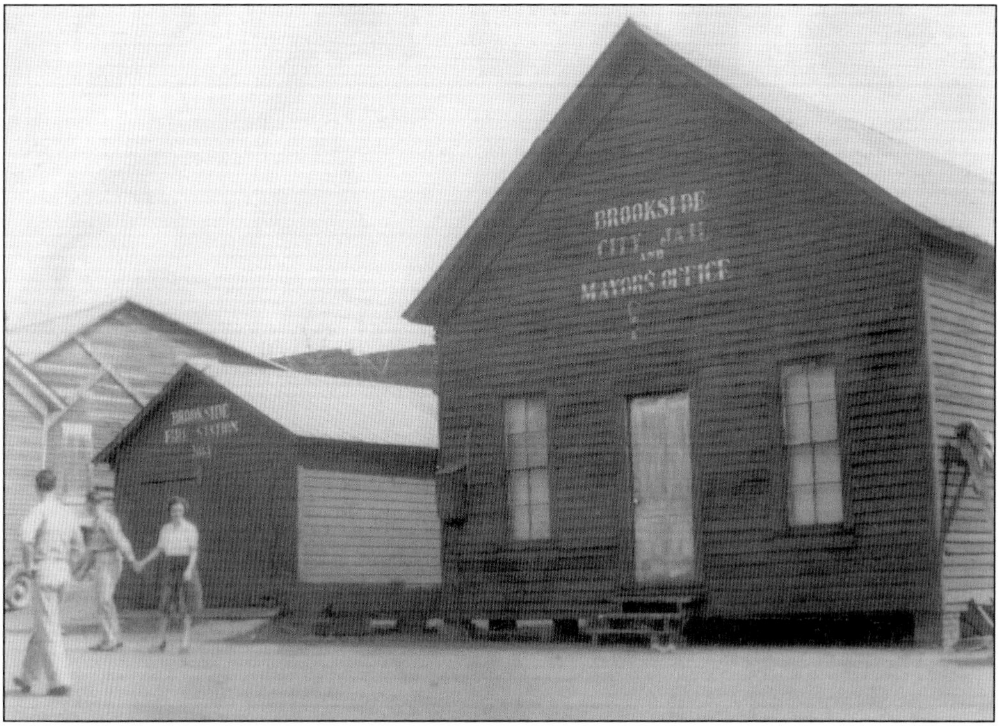

Company towns were carved out of the Jefferson County hill country, replacing small farms and undisturbed wilderness. Many coal towns earned reputations for violence, and the popular literature of the day only confirmed the stereotype. In *Black Earth*, Tom Rowan described Brookside as a hard scrabble half-cocked type of town. (BHM.)

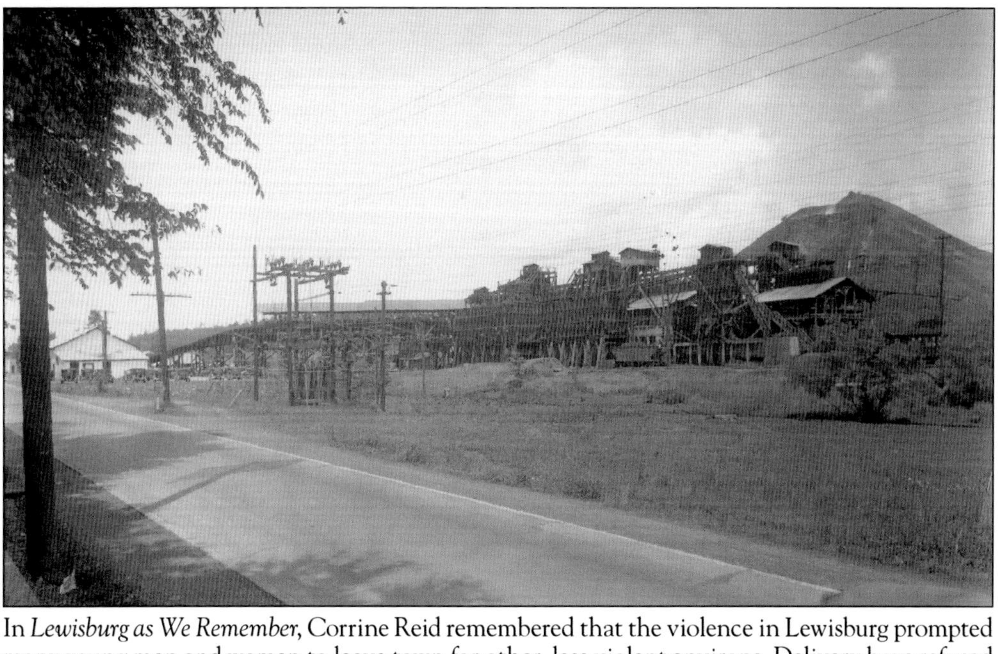

In *Lewisburg as We Remember*, Corrine Reid remembered that the violence in Lewisburg prompted many young men and women to leave town for other, less violent environs. Delivery boys refused to deliver parcels to a section of Lewisburg known as Bloody Beat. (BPL Collection.)

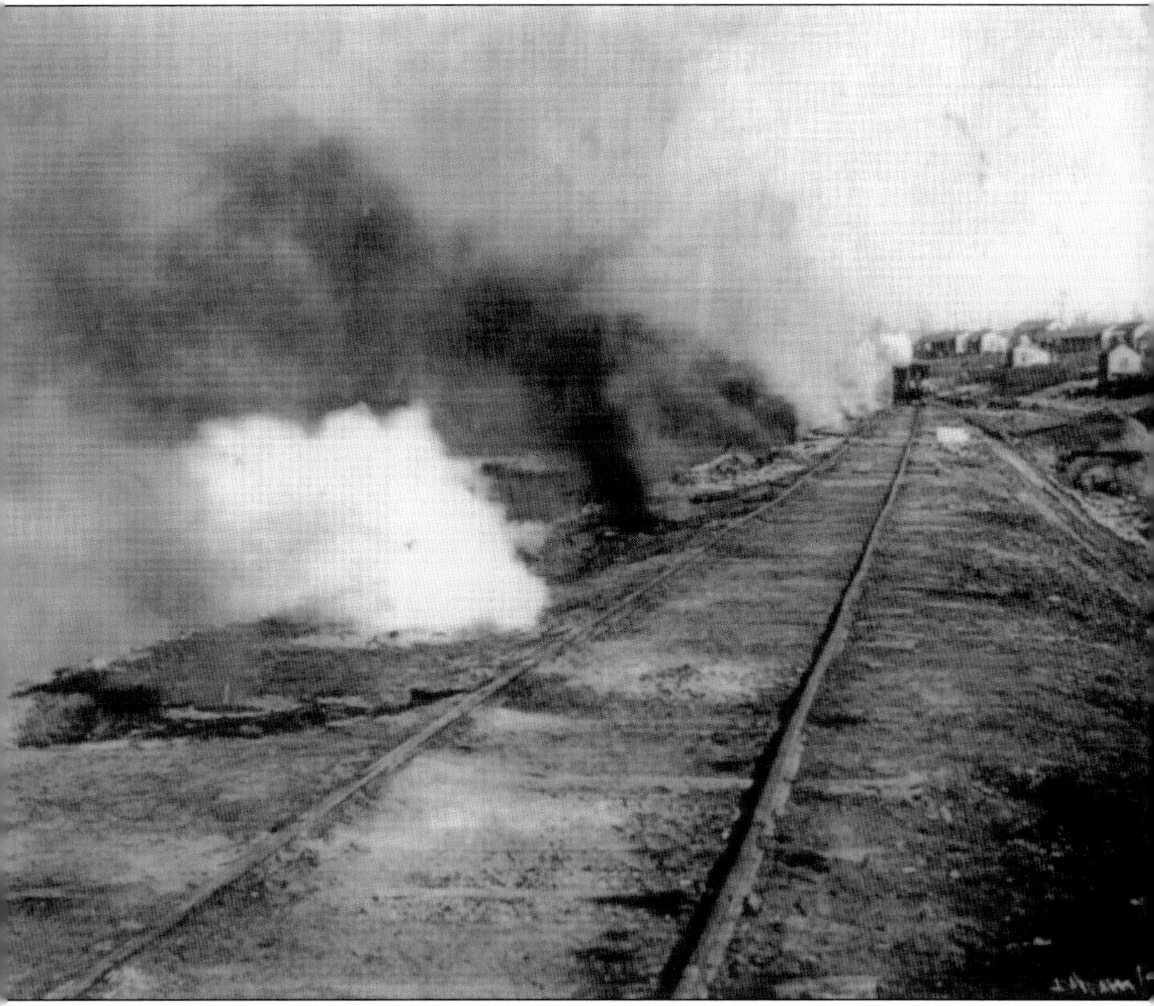

Fultondale grew into being when Fulton Springs and Glendale merged in the early 20th century. Lewisburg was located between Fulton Springs and Birmingham. The mines began to close mid-century. (BPL Collection.)

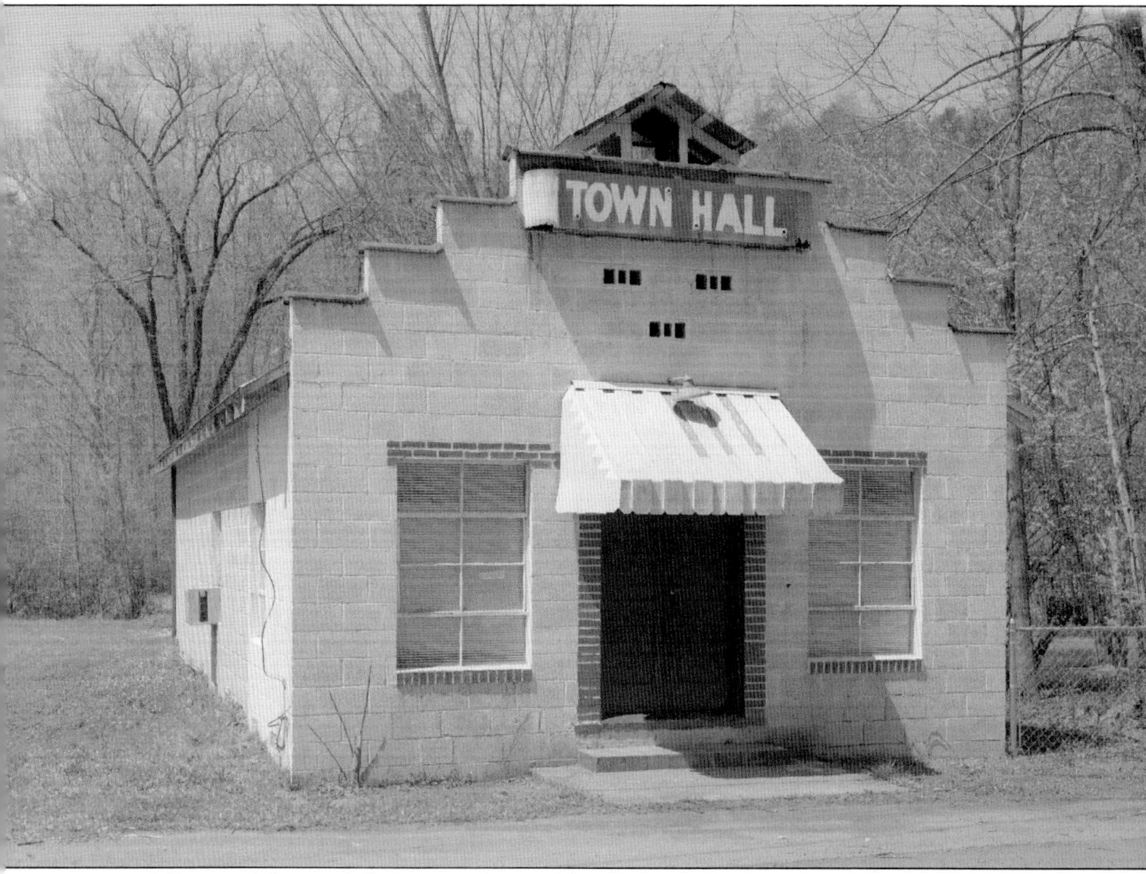

Cardiff, named for Cardiff, Wales, is located between Brookside and Graysville and, during her heyday from 1900 to 1920, attracted French, Welsh, Italian, and English immigrants. After several fires destroyed much of the town's center and the 1920 coal strike effectively closed the mines, Cardiff transformed into a small, sleepy town. (LOC, HABS.)

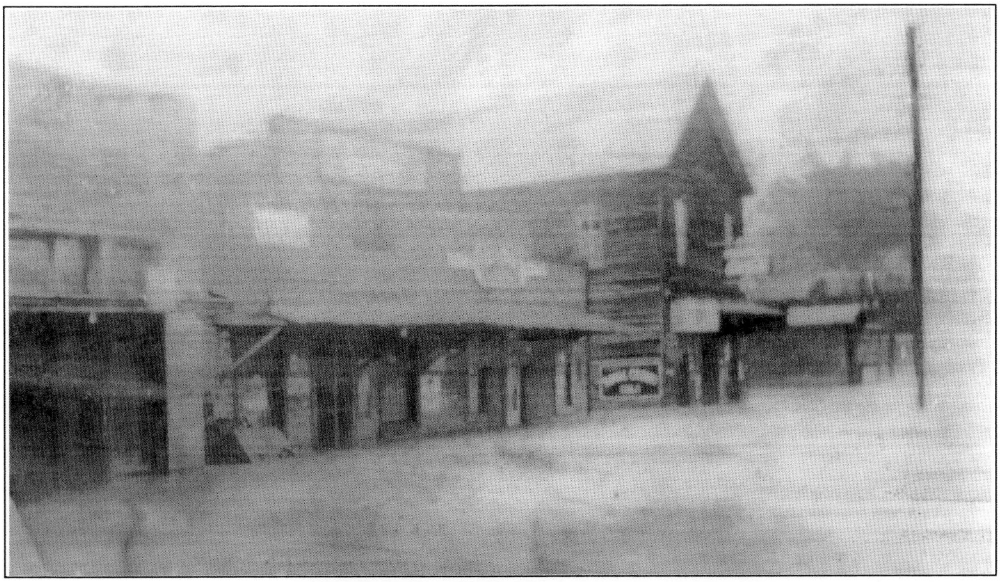

Most mining communities had a main street filled with stores, from company-owned to privately owned. The types of stores ran the gamut from dress shops to saloons. Some, like H. Russel's Saloon and Tony's on the Bridge in Brookside, issued trade tokens for use in their shops. (BHM.)

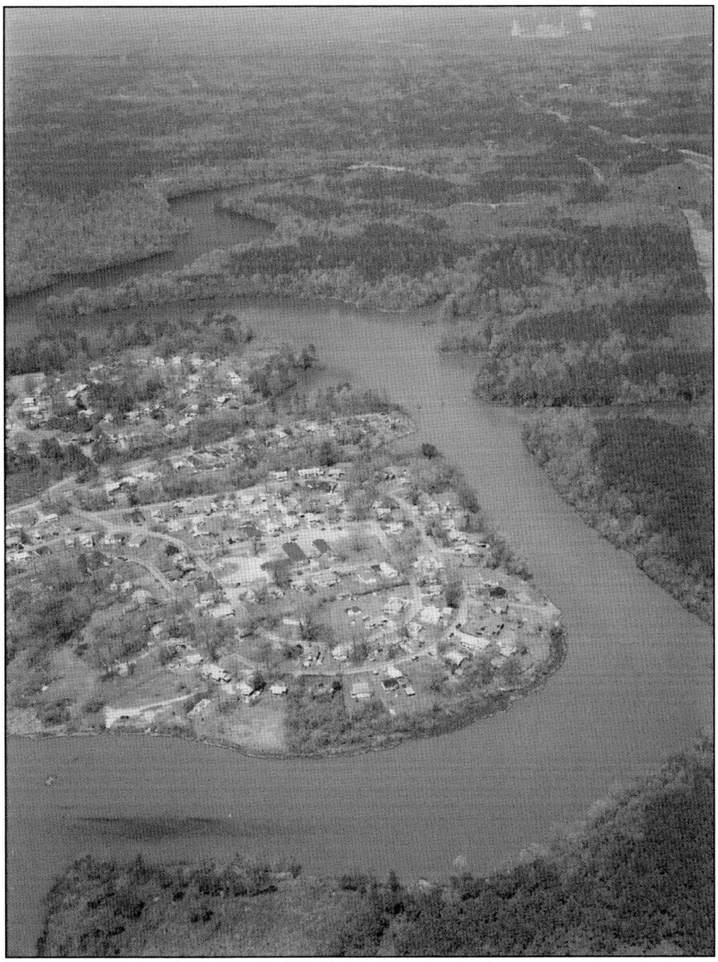

Bayview was one of TCI's model villages and, in fact, won an award at the New York World's Fair. Its curving shape, pleasingly designed roads, houses, and community facilities made Bayview one of the jewels in the TCI crown. It even had garbage service operated by the company. (LOC, HABS.)

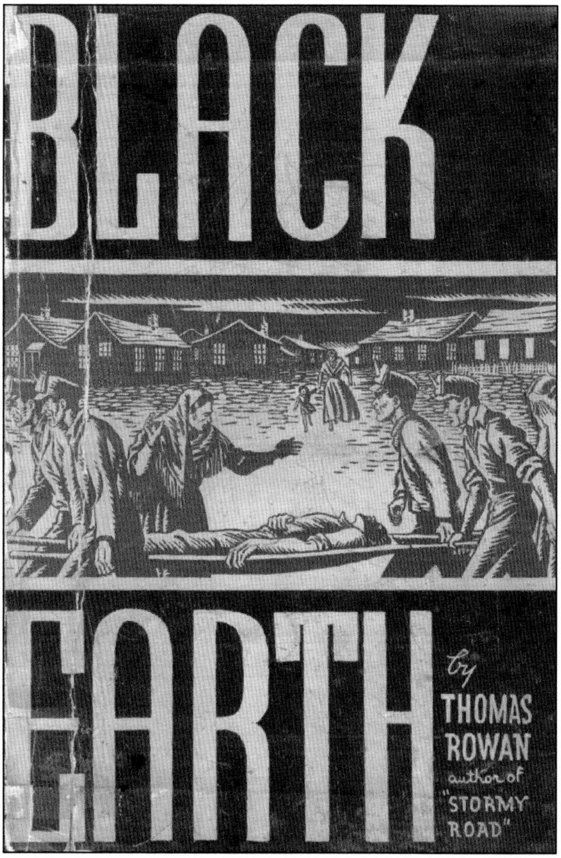

Brookside native son Thomas Rowan made a splash with his novel *Black Earth* in the mid-1930s. He described Cardiff, Banner, and Brookside and the surrounding mining communities as something wild and primitive with labor strife, hard living, hard drinking, and hard playing. (Author's collection.)

Settled in a picturesque farming valley, Republic was a coal town made of three different sections. One section was named Early Town, another named New Town, and yet another named East Warner. Republic school originally went through grade 12, but as the mines closed and the population shrank, it became an elementary school. (BPL Collection.)

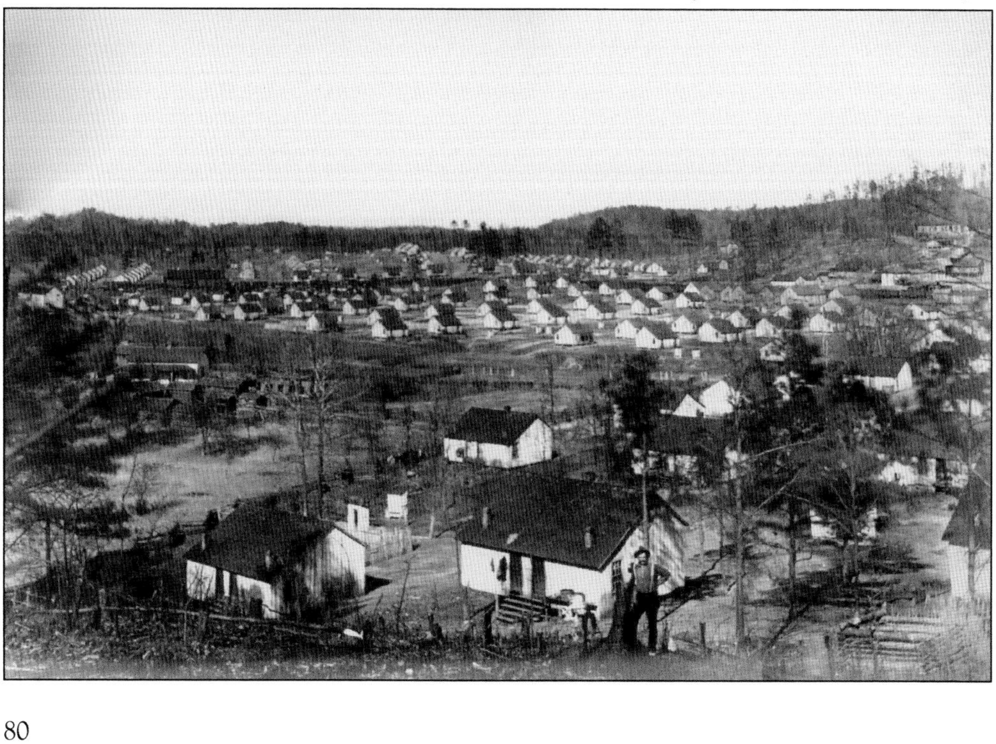

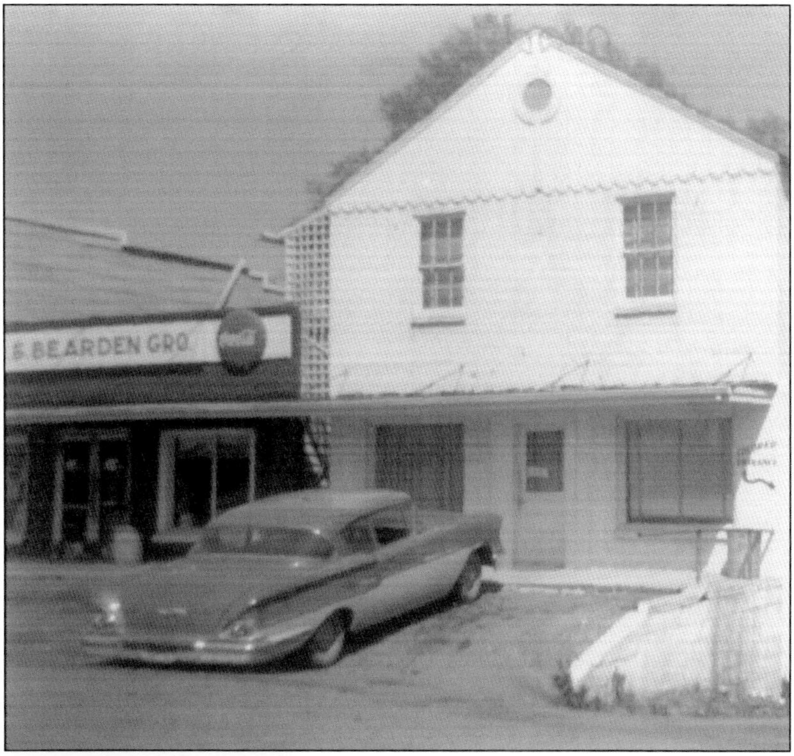

The mines at Gintown, Graysville today, were operated by Sloss-Sheffield Iron and Steel Company. In 1920, at the worst of the nationwide coal strikes, a union miner was charged with the murder of a nonunion miner. (BPL Collection.)

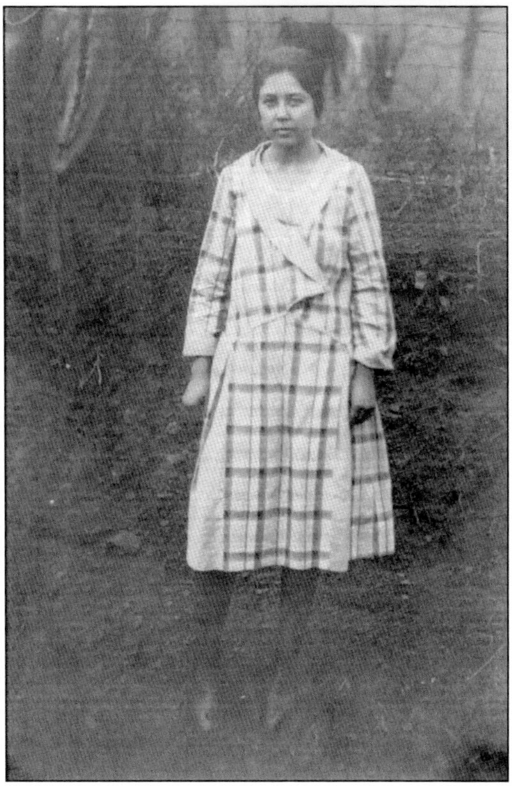

Blossburg was a mining town located between Pinckney City and Gin Town, or Graysville. It had been the scene of labor unrest and many of the mines were operated by the Brookside-Pratt Company. Addie Black, pictured here, was one of the store clerks at the old commissary. At age 16, she bought her first doll from the commissary with her earnings. (Author's collection.)

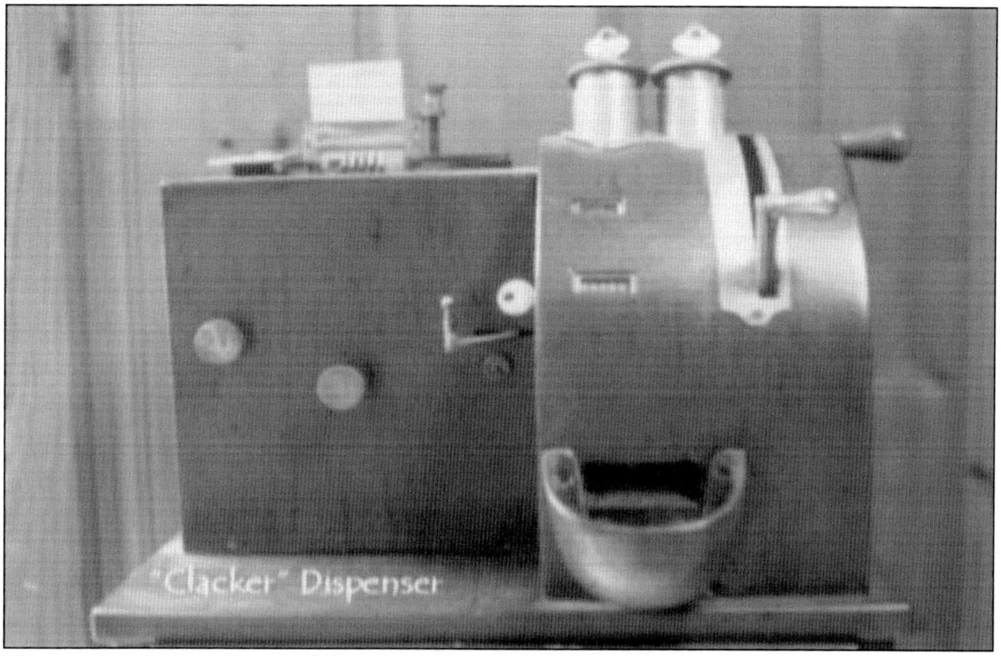

The company store was one of the central parts of many coal mining families' lives. Each town had one, and many miners were obligated to buy from them since the company paid them in scrip or clacker. Scrip is company-issued paper money, while clacker was company-issued brass tokens with designations from 1¢ to $1. (Defore.)

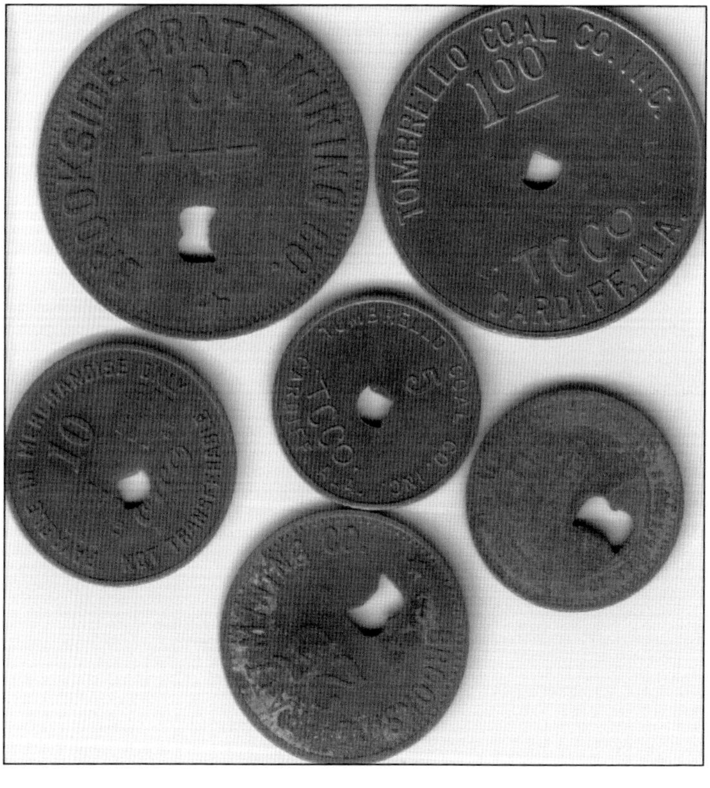

Some Slovaks in Brookside refused to take all of their pay in clacker, demanding cash instead. Albert Billits, who worked at the Brookside-Pratt Commissary, reported that many in the community bought clacker from miners so that they could shop at the commissary, as the store had some of the best selections of merchandise. (Author's collection.)

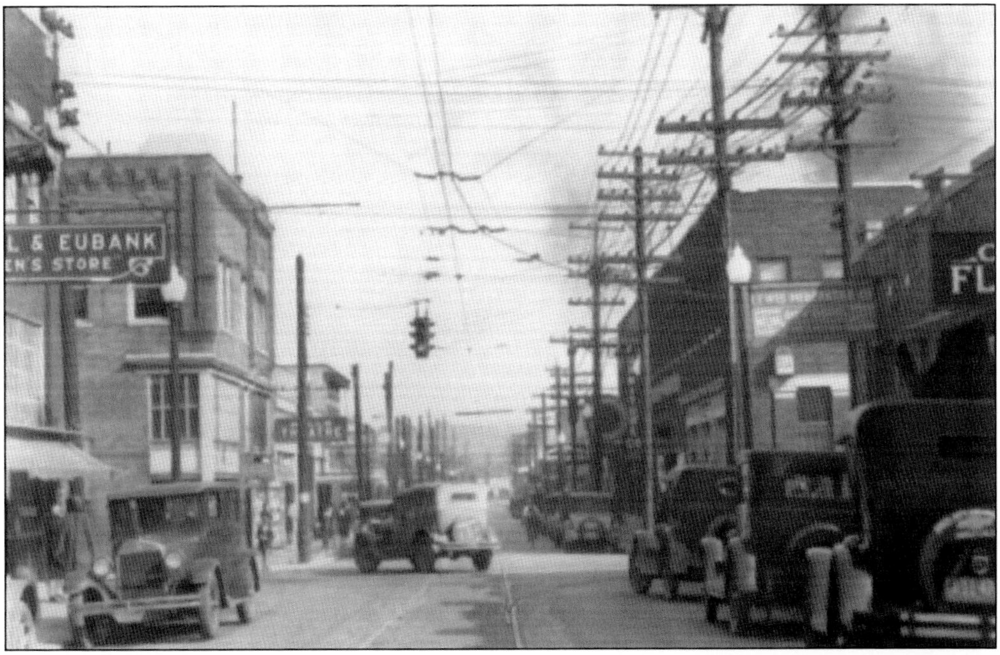

Ensley was one of the more thriving company towns to grow up around the iron and coal industries. Founded in 1886 by Enoch Ensley, the city was one of the first to employ modern systems of water and sanitation control. TCI bought a controlling interest in Ensley's company and built their first furnace and steel mill in Alabama here. (BPL Collection.)

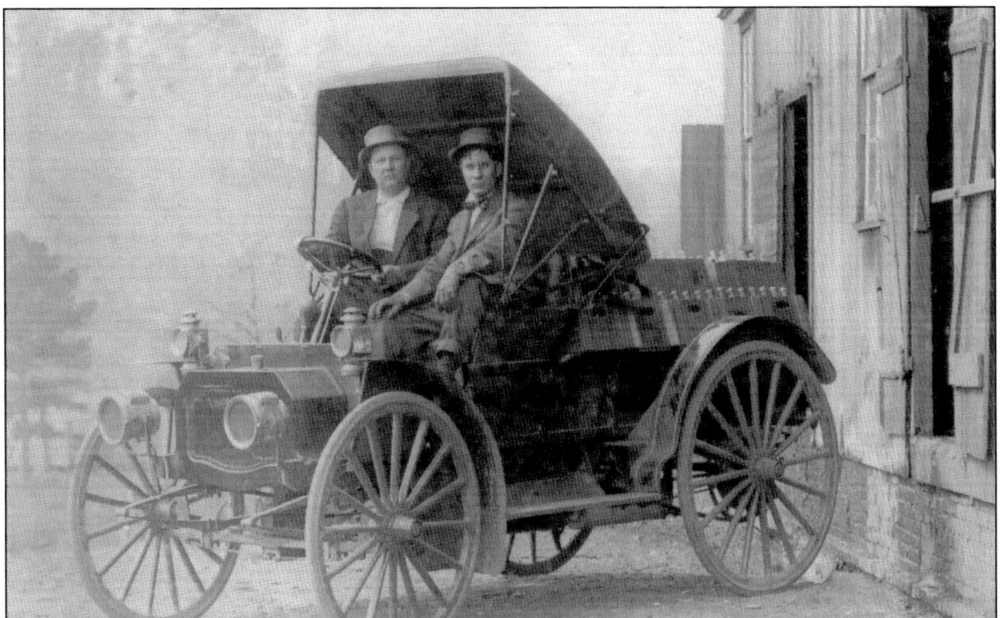

Many of the early mining towns had their own soda bottling facilities. George Park Davis owned a franchise of Pioneer Bottling Works in Brookside and in Blossburg. Most towns had their own local soda brand that played on the Coca-Cola brand. For Brookside, Coca-Lula was the favorite. It was a source of pride for the town. (BHM.)

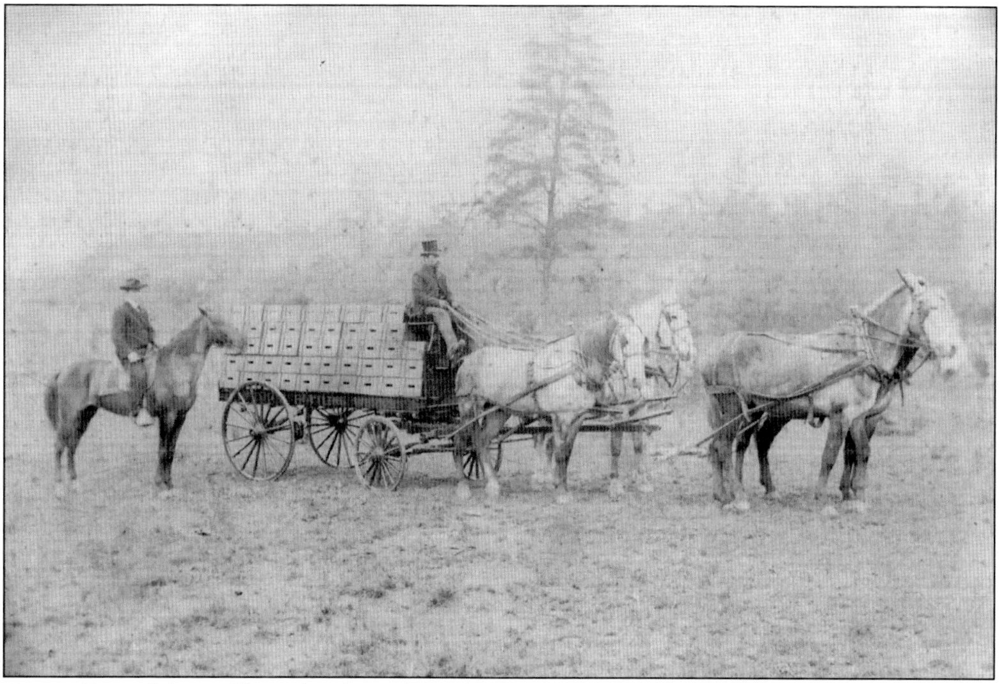

George Davis delivered his colas throughout the county's western mining camps. His own formula proved so popular that Coca-Cola bought it to keep him from competing with them. (BHM.)

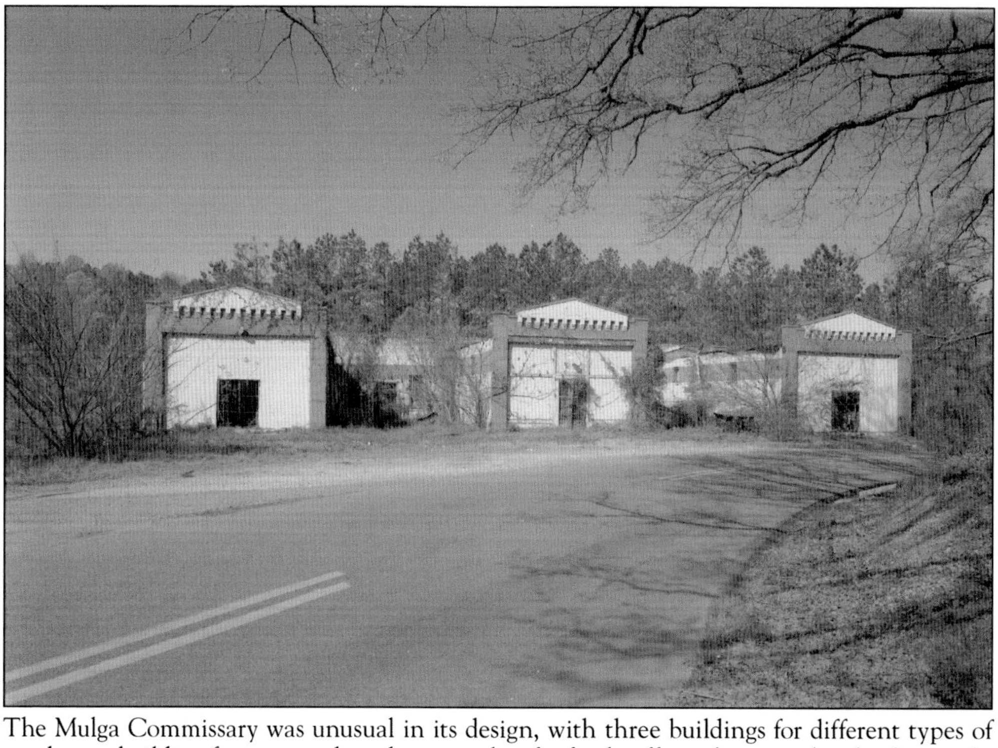

The Mulga Commissary was unusual in its design, with three buildings for different types of goods: one building for meat and produce, another for foodstuffs, and yet another for dry goods. (LOC, HABS.)

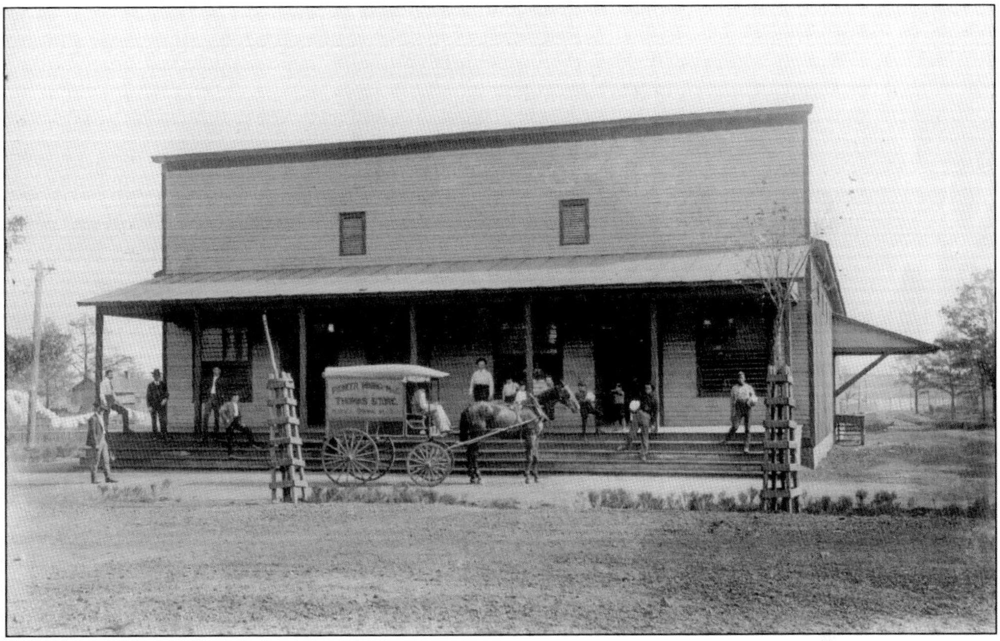

The Pioneer Mining Company at East Thomas served residents who lived in neat row houses nearby. The East Thomas community was an urban mining town and its buildings and streets reflected the town's urban environs. (BPL Collection.)

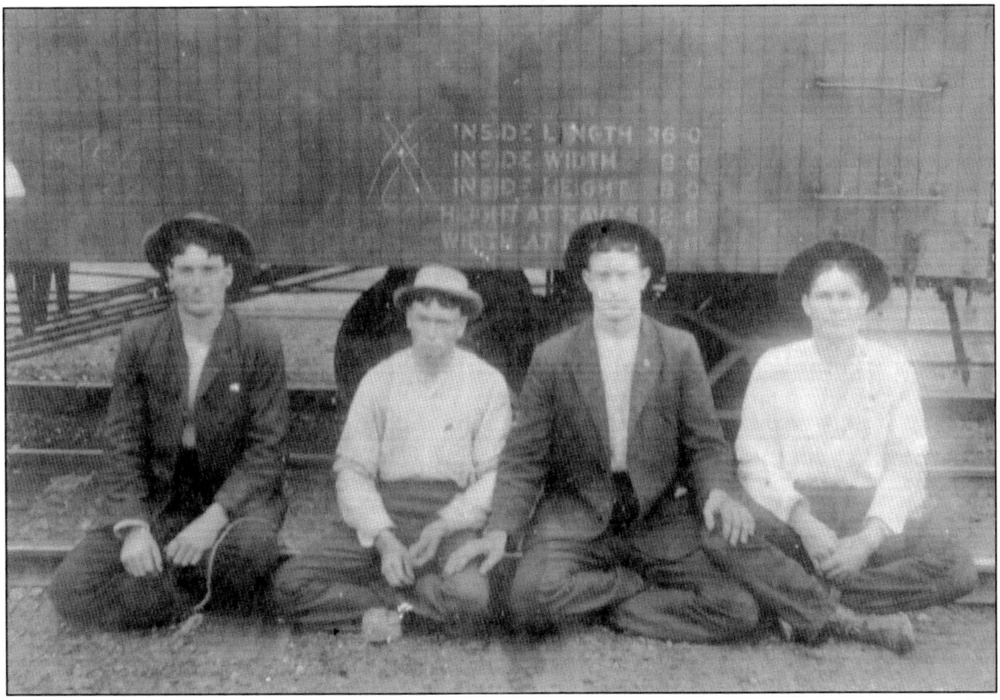

These young men spend a carefree moment against a boxcar outside the depot at Watson. Supplies to the stores of the coal towns were delivered by train. Brookside mayor Roger McCondichie remembered waiting at Brookside's depot for a basketball that he had ordered to be delivered. (Author's collection.)

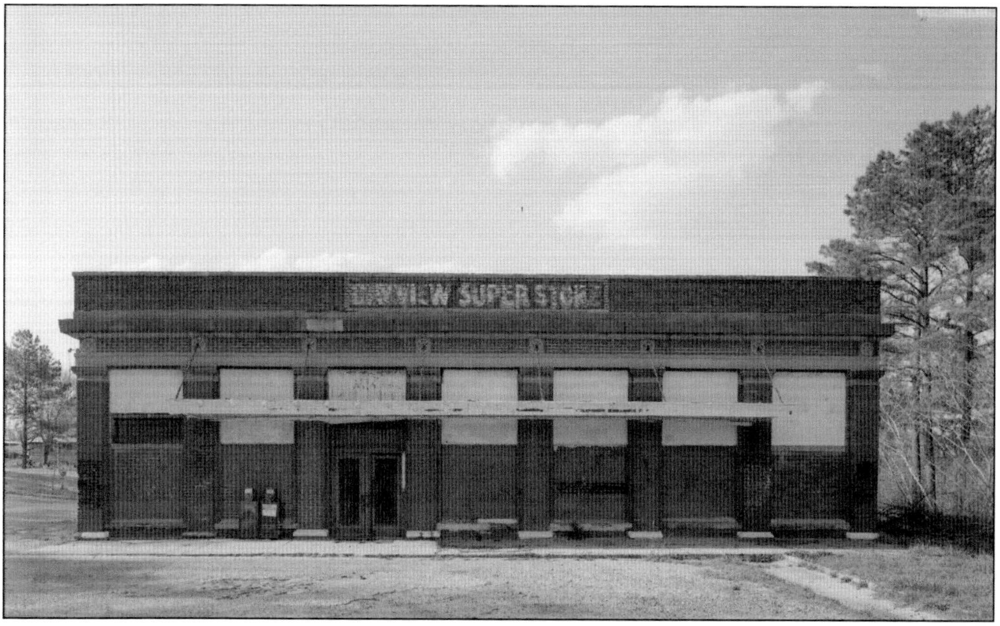

Bayview was another of TCI's planned communities. It was a companion to Docena and Edgewater. It was complete with full-size commissary, schools, and recreational centers. (LOC, HABS.)

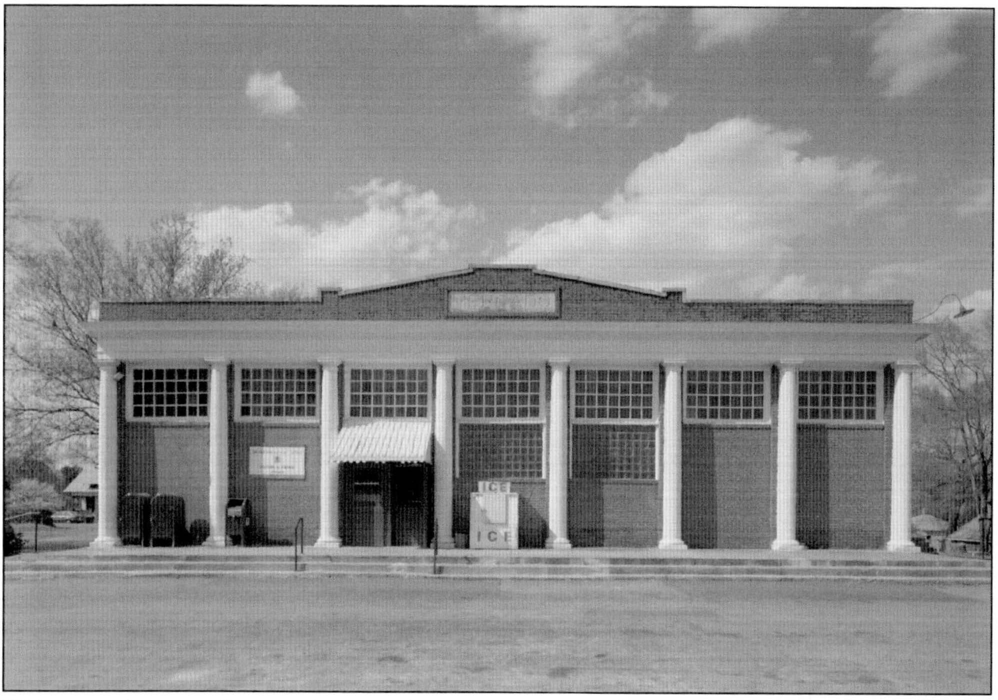

This commissary at Docena is unusual in comparison with many of the other store designs. While many of the other company stores are plain brick buildings created for function over style, this commissary reflects a stately, more pleasing design. Docena was one of TCI's model towns. (LOC, HABS.)

This is a pay stub for Albert Billits, who worked as a clerk at the Brookside Pratt Commissary in Brookside. (BHM, Albert Billits.)

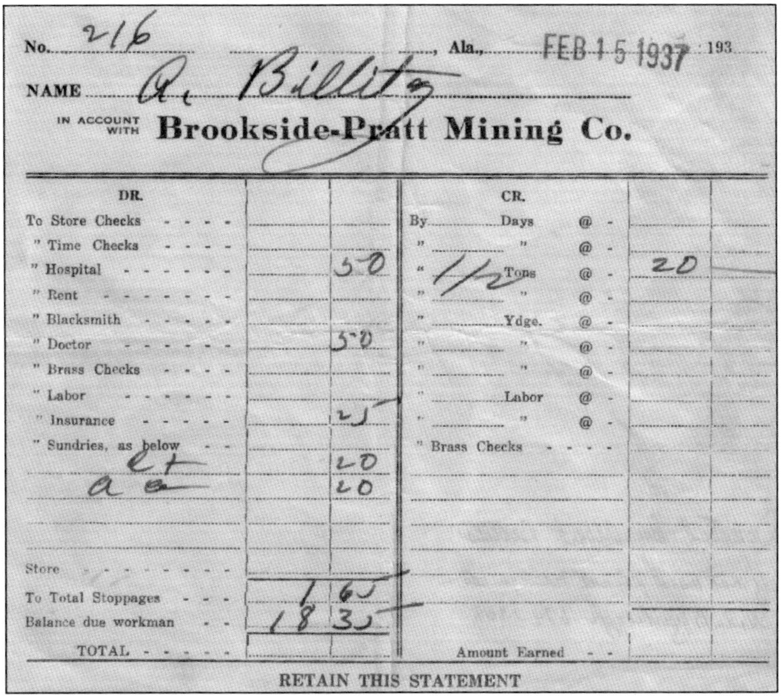
No. 216 ...., Ala., FEB 15 1937 193

NAME A. Billits

IN ACCOUNT WITH **Brookside-Pratt Mining Co.**

DR.			CR.		
To Store Checks - - - -			By .... Days @ -		
" Time Checks - - - -			" .... " @ -		
" Hospital - - - - - -		50	" 1/2 Tons @ -	20	
" Rent - - - - - - -			" .... " @ -		
" Blacksmith - - - - -			" .... Ydge. @ -		
" Doctor - - - - - -		50	" .... " @ -		
" Brass Checks - - - -			" .... " @ -		
" Labor - - - - - -			" .... Labor @ -		
" Insurance - - - - -		25	" .... " @ -		
" Sundries, as below - -			" Brass Checks - - - -		
et		20			
a a		20			
Store - - - - - - - -					
To Total Stoppages - - -	1	65			
Balance due workman - -	18	35			
TOTAL - - - - -			Amount Earned - -		

RETAIN THIS STATEMENT

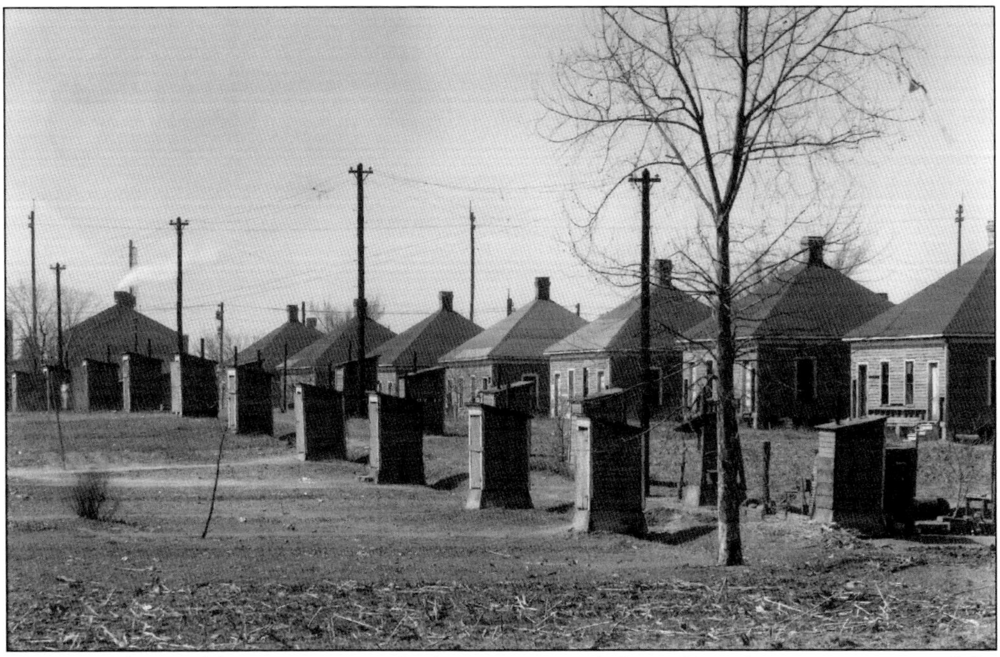
The Republic Steel houses are typical of many scattered throughout the mining camps of the district. (LOC.)

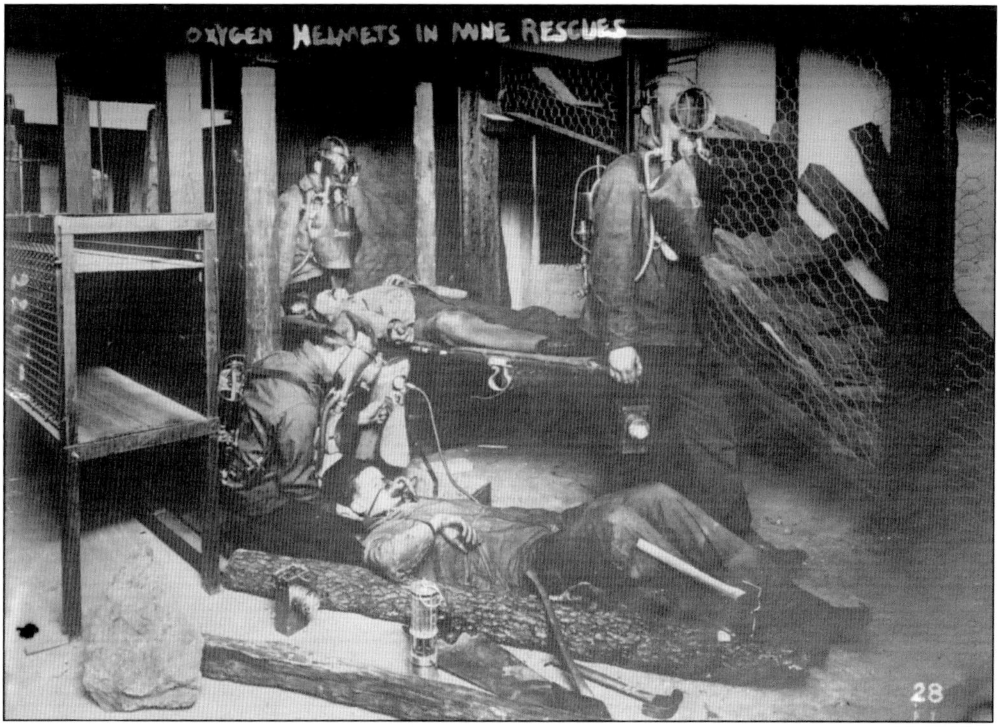

The Praco community was rocked by disaster in 1938 when a rock wall gave way, burying some men and injuring others. One man's arm was amputated to remove it from a rock, and a mine superintendent was crushed by falling debris. Hampering the rescue was the depth of the shaft, at 3,000 feet. (LOC.)

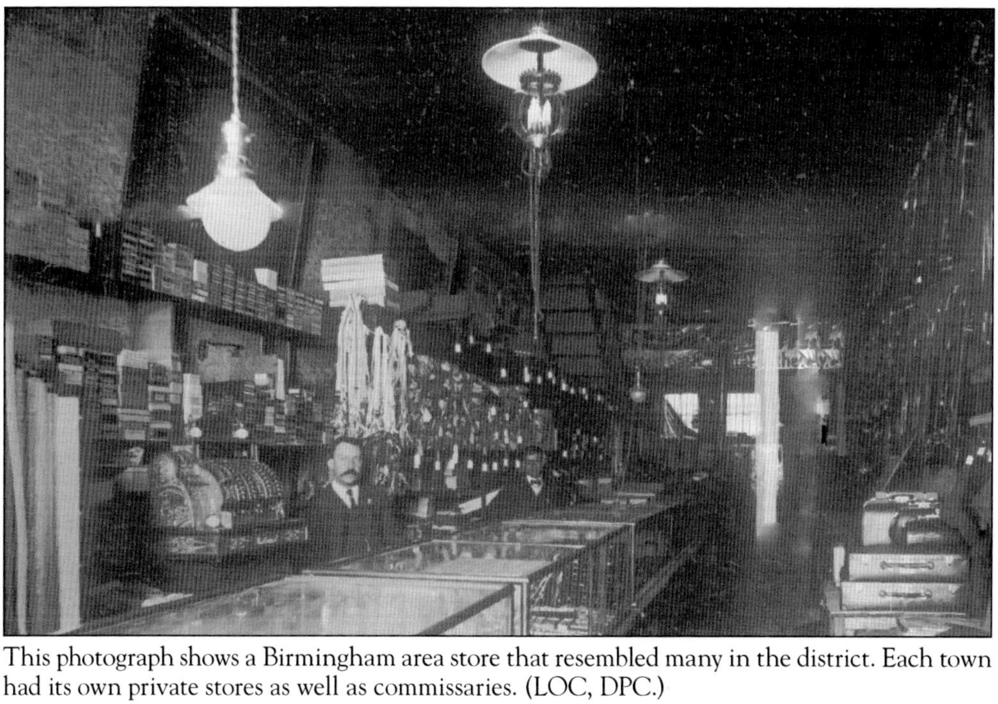

This photograph shows a Birmingham area store that resembled many in the district. Each town had its own private stores as well as commissaries. (LOC, DPC.)

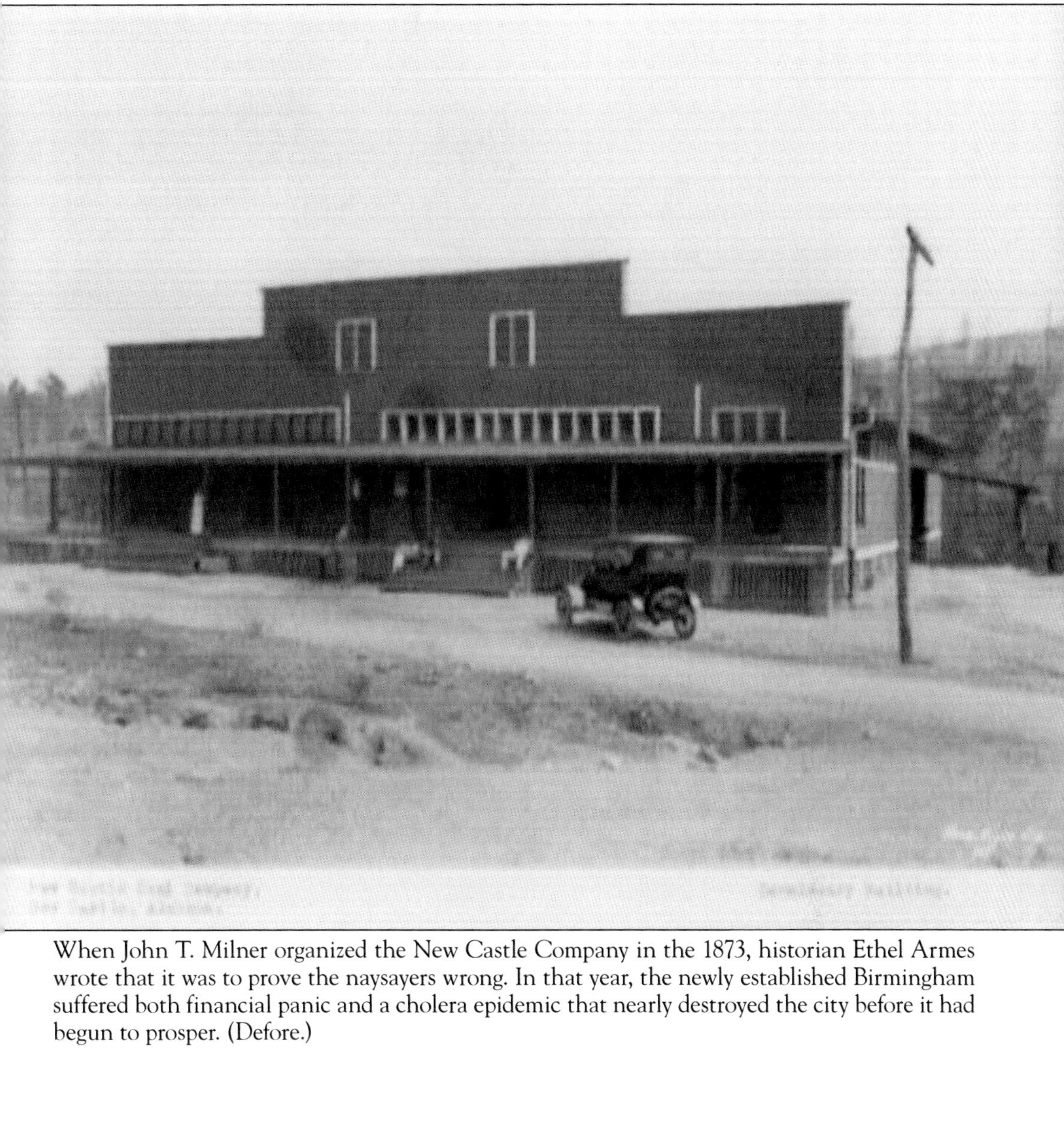

When John T. Milner organized the New Castle Company in the 1873, historian Ethel Armes wrote that it was to prove the naysayers wrong. In that year, the newly established Birmingham suffered both financial panic and a cholera epidemic that nearly destroyed the city before it had begun to prosper. (Defore.)

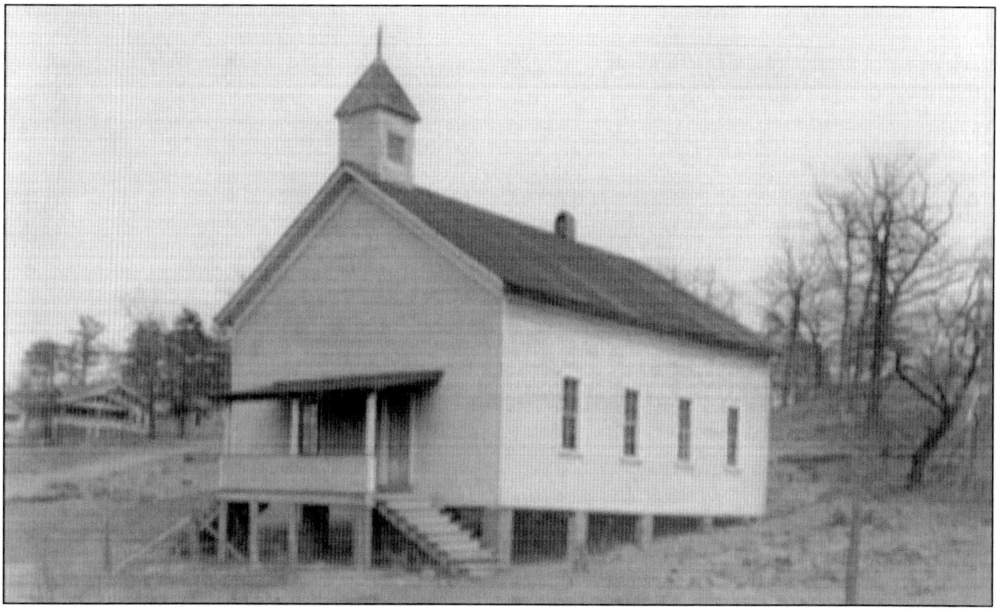

John Milner had been responsible for building railroads through the state, notably the North and South line, and was widely recognized as a "father" of Birmingham and the industrial district that surrounded it. Milner built facilities for both black and white workers. The above is the black church at New Castle. (Defore.)

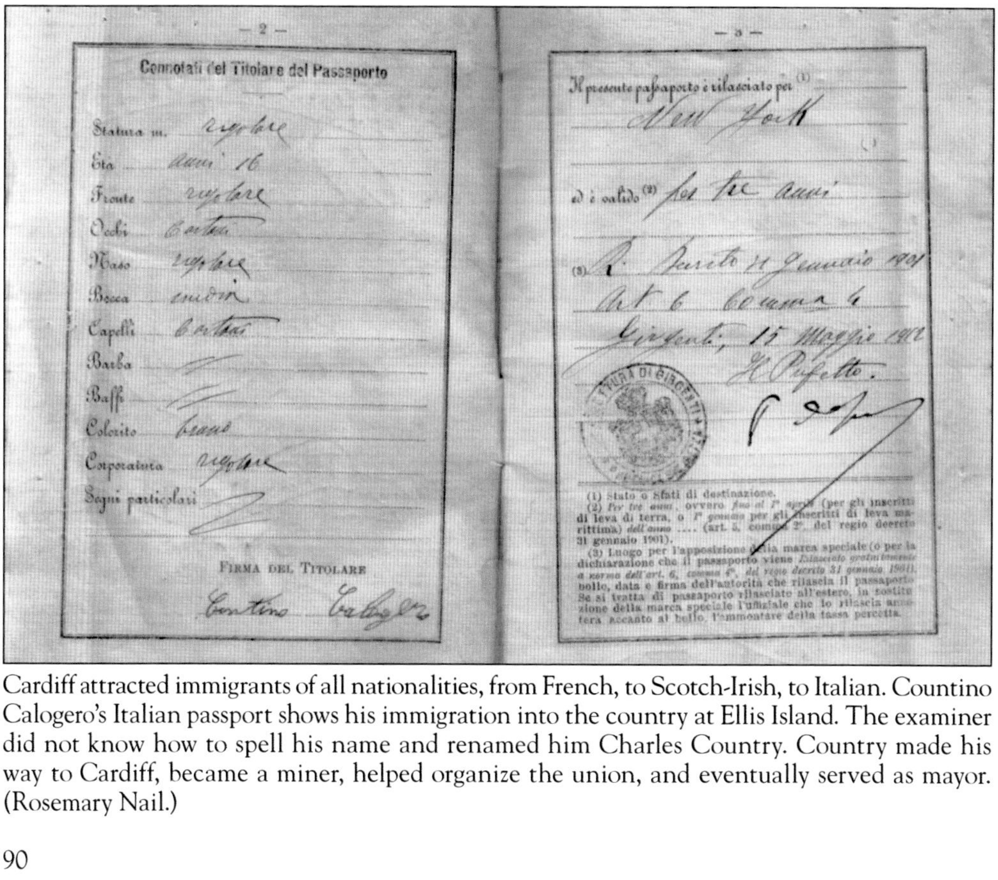

Cardiff attracted immigrants of all nationalities, from French, to Scotch-Irish, to Italian. Countino Calogero's Italian passport shows his immigration into the country at Ellis Island. The examiner did not know how to spell his name and renamed him Charles Country. Country made his way to Cardiff, became a miner, helped organize the union, and eventually served as mayor. (Rosemary Nail.)

This map shows the industrial railroads that linked many of the coal mining towns together with each other and the furnaces that they supplied. It shows a collection of the most important TCI sites and model communities: Mulga, Bayview, and Docena. Republic's Sayreton mines are located to the east of Birmingham. (LOC, HABS.)

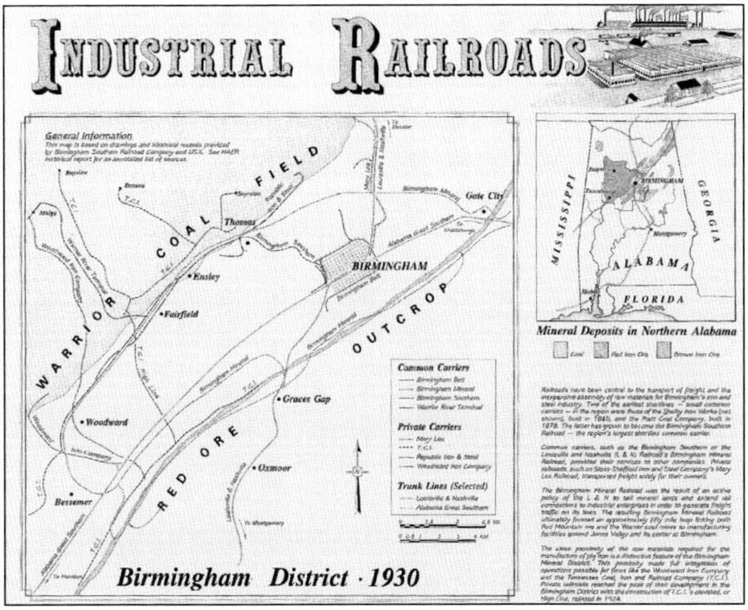

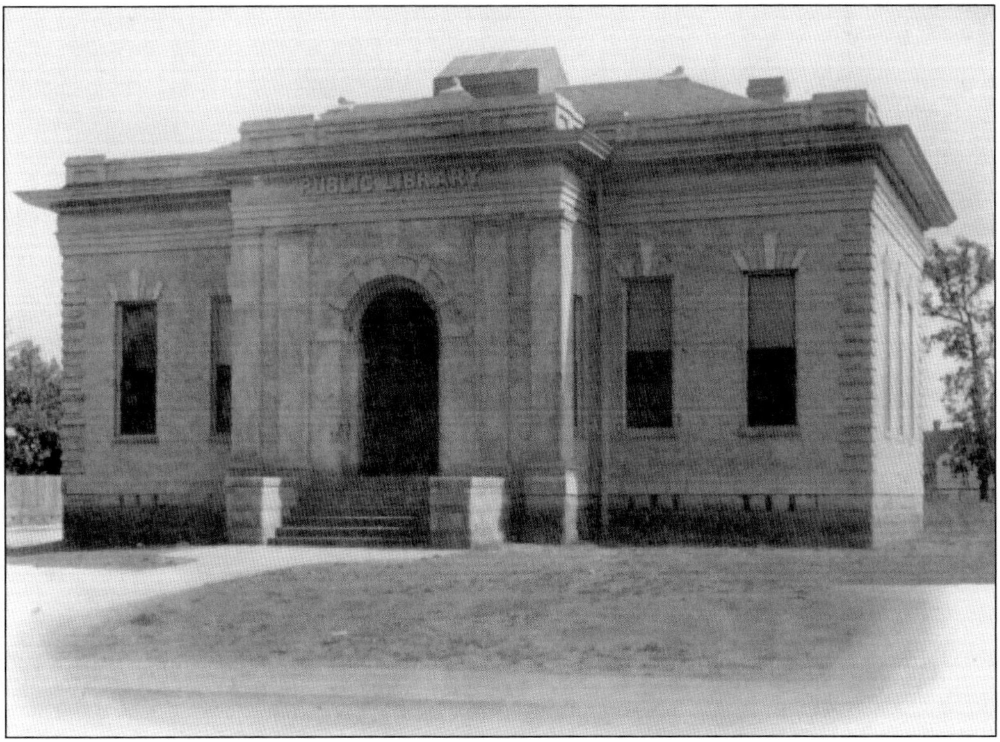

Ensley, a mining and steel worker community fueled by the TCI furnaces, was one of 14 communities to apply for and receive monies by the Carnegie Foundation for the building of a public library. Ensley had its own city government, school system, and streetcar line before being incorporated into Birmingham in 1910. (LOC, DPC.)

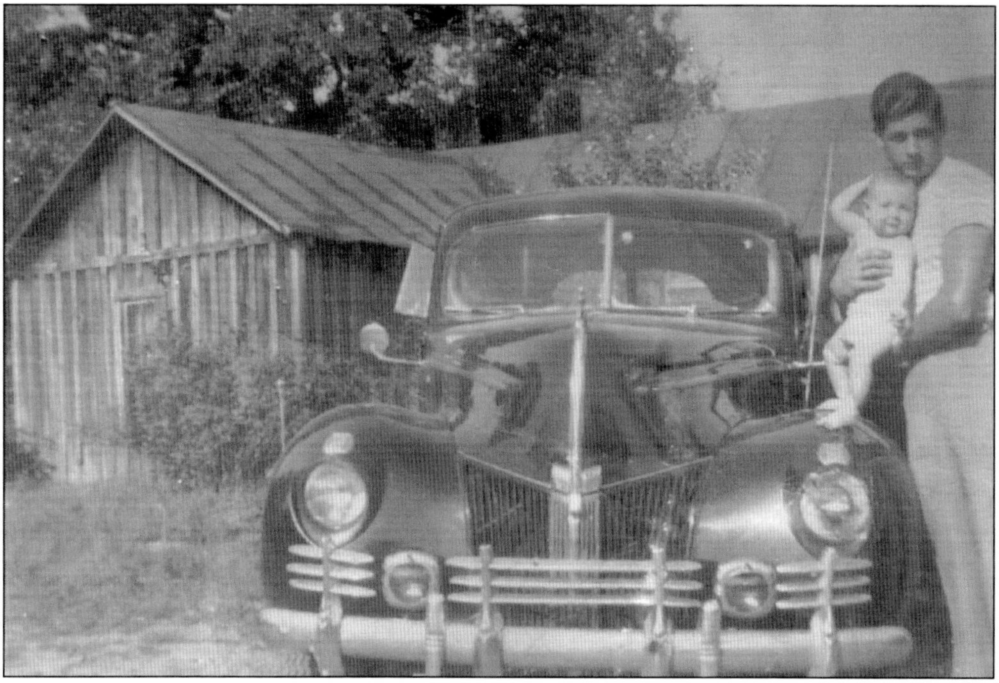

This photograph shows the typical style of house in many coal towns. This company house had three rooms, no running water, and an outhouse. Its walls were paper thin, allowing in the cold wind on a winter's night. (Author's collection.)

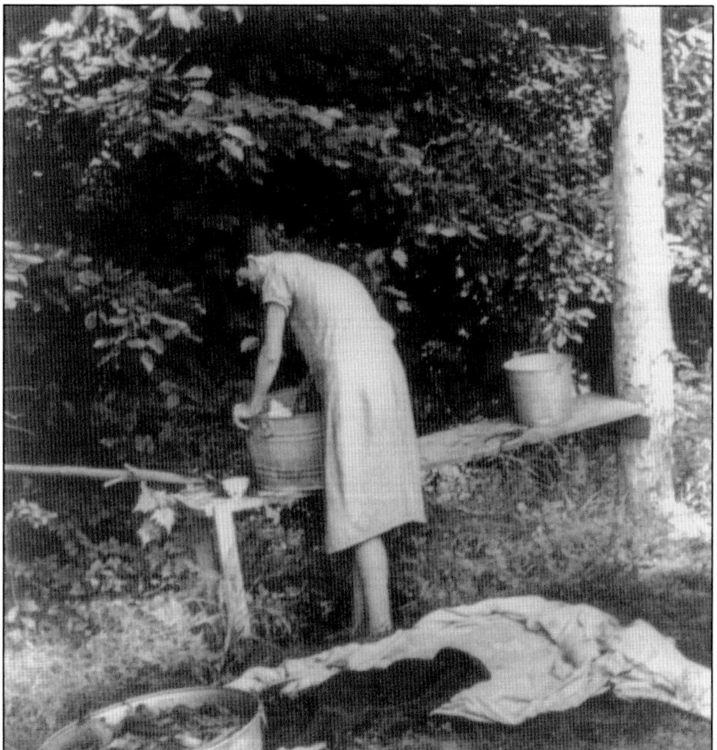

While mining families lived in towns with stores and some modern amenities, many still did things the old-fashioned way. Ralph Sanford remembered Mondays being wash day at his Republic home. His mother made lye soap from hog fat and then washed her clothes over an open fire. (LOC, FSA.)

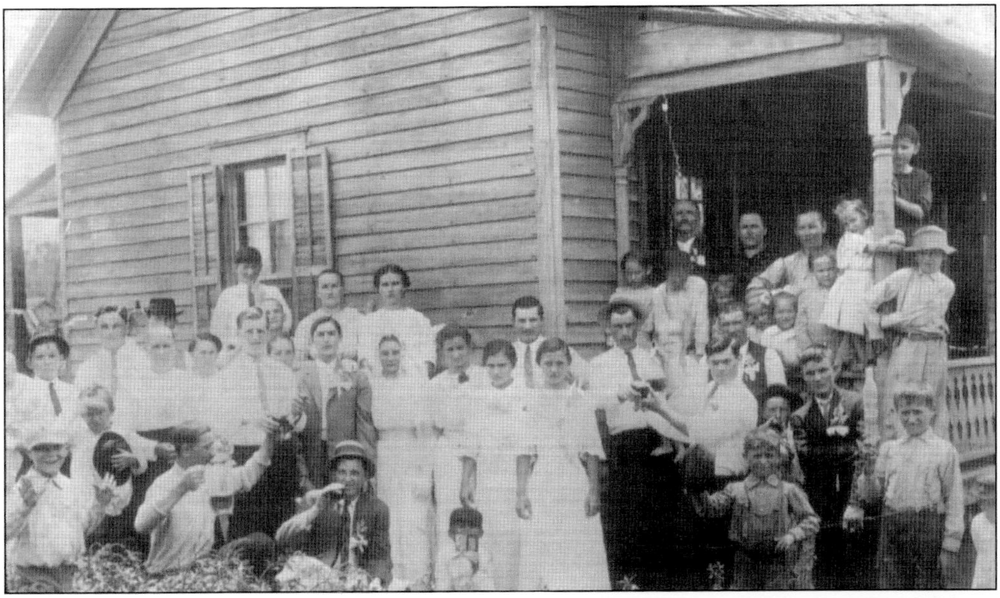

This home was the parish house for SS Cyril and Methodius Catholic Church in Brookside. It is, however, indicative of the private homes in many coal towns. Brookside was divided into the company camp and the incorporated town with private homes that were larger and well kept. (BHM, George Slovensky.)

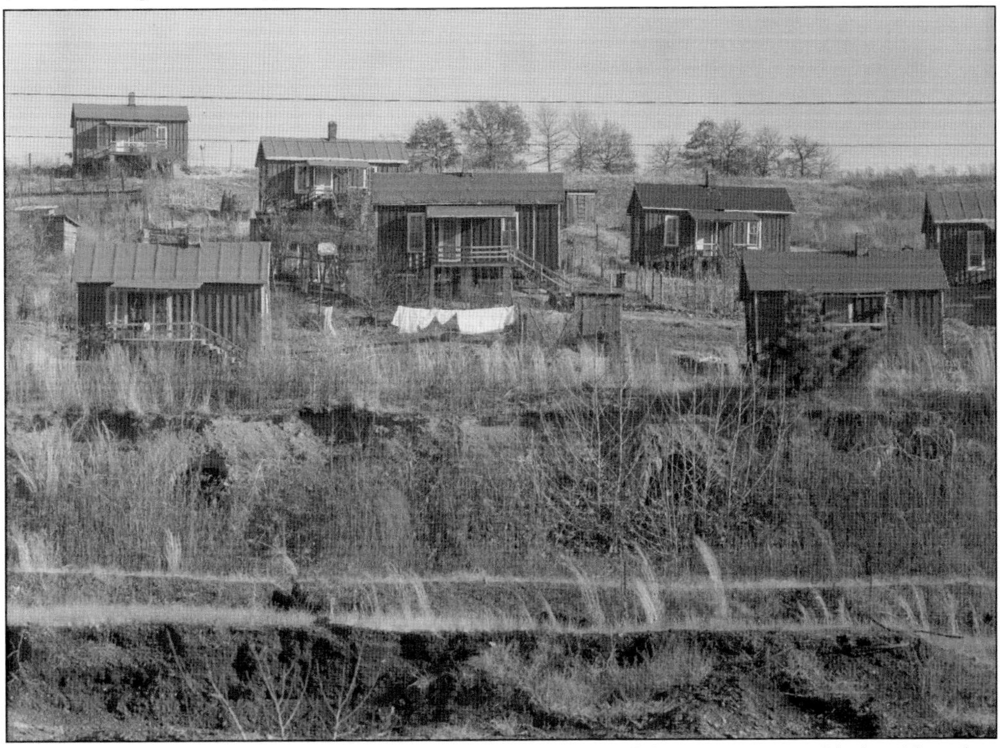

Unfortunately, many of the coal towns in the rural parts of the county were blighted by what outsiders called slums. These were no more than shacks interspersed between the rolling hills and scraggly pines of the county's hill country. (LOC, FSA.)

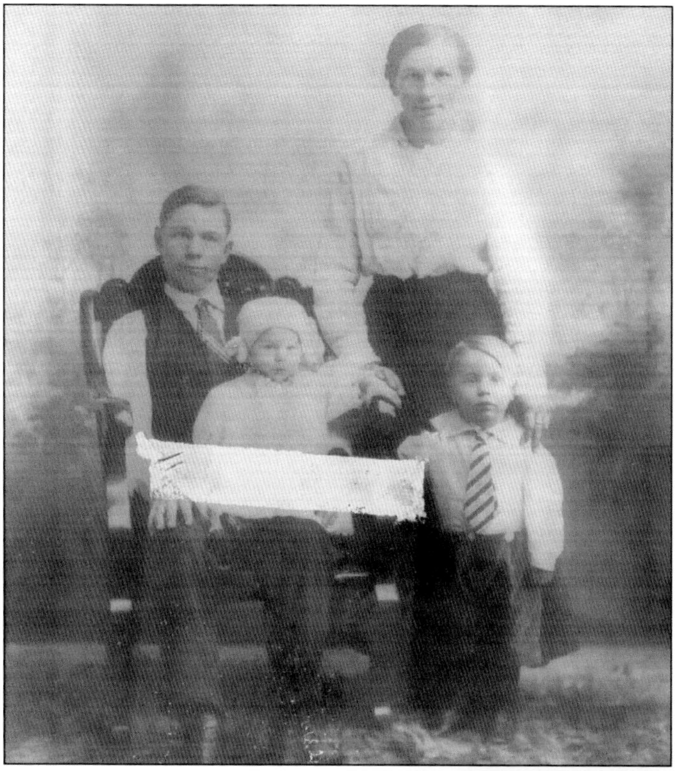

The closing of the mines left many with few job opportunities. Many like this family left for northern states where jobs were available. (BHM.)

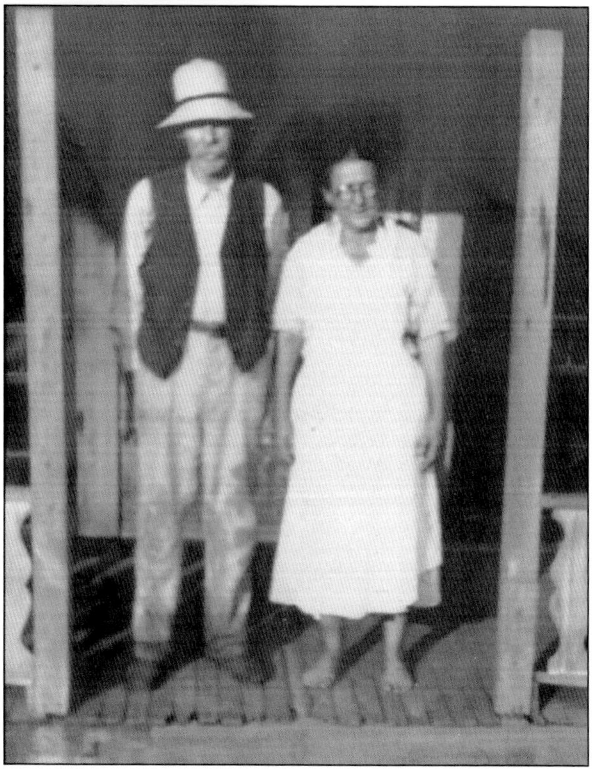

This couple, J.T. and Epsie Black, lived in Pinckney City near Blossburg. When the mines closed in the 1920s, many of their sons and daughters moved with their families to Illinois to find work. (Author's collection.)

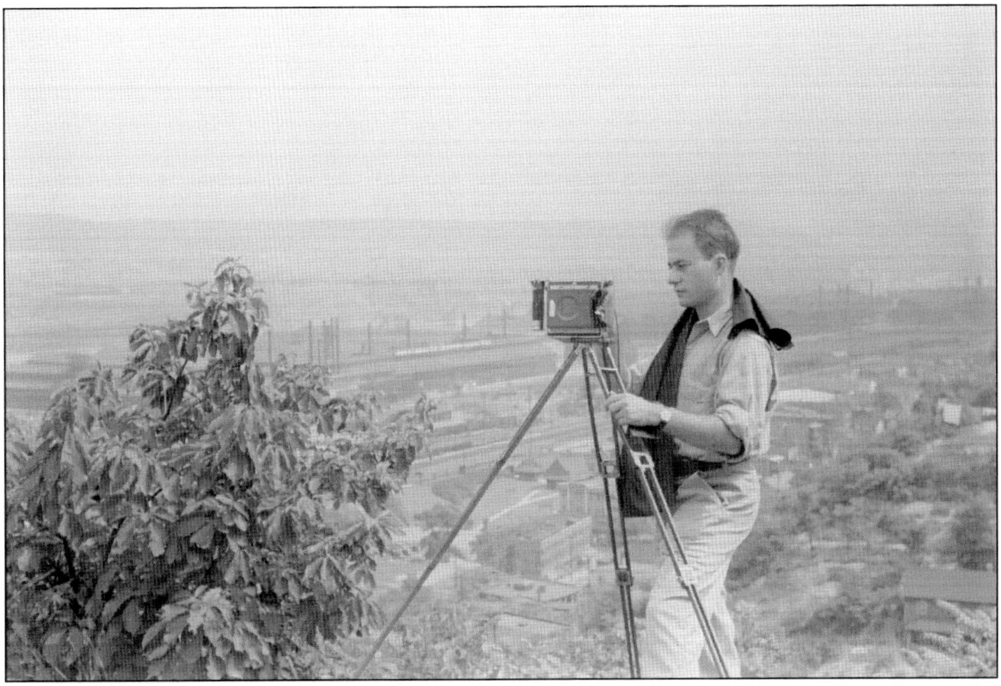

Arthur Rothstein was one of the most famous of the US government photographers during the period from 1935 throughout the 1940s. He crisscrossed the South and is primarily responsible for documenting many of the area's coal miners and their living and working conditions. (LOC, FSA.)

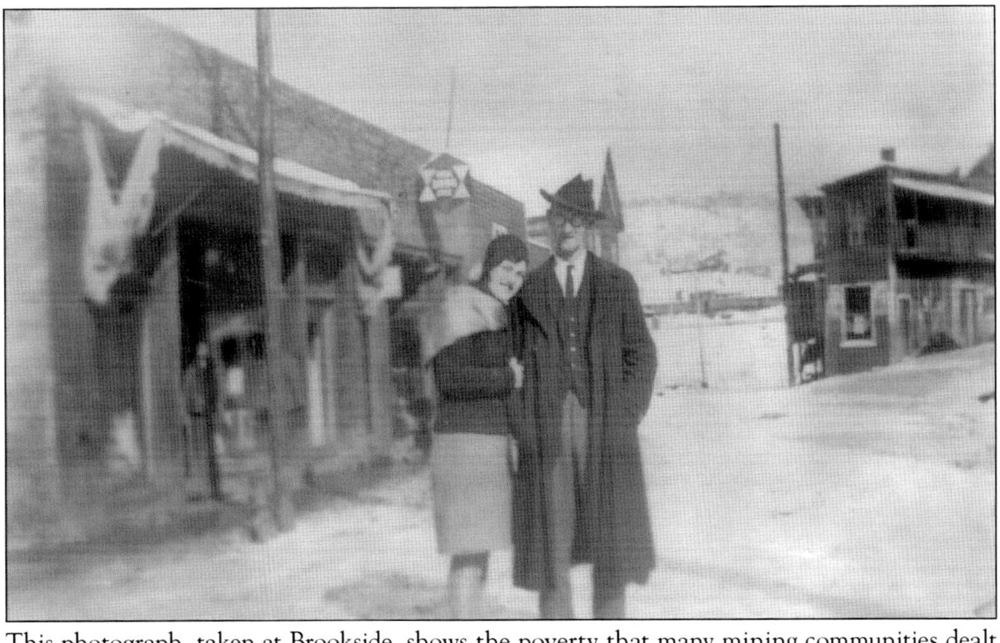

This photograph, taken at Brookside, shows the poverty that many mining communities dealt with after the major mines closed. Downtowns were tattered and worn looking and displaced workers had little money for repairs to their own homes. Some resorted to making illegal alcohol to put food on the table. (BHM.)

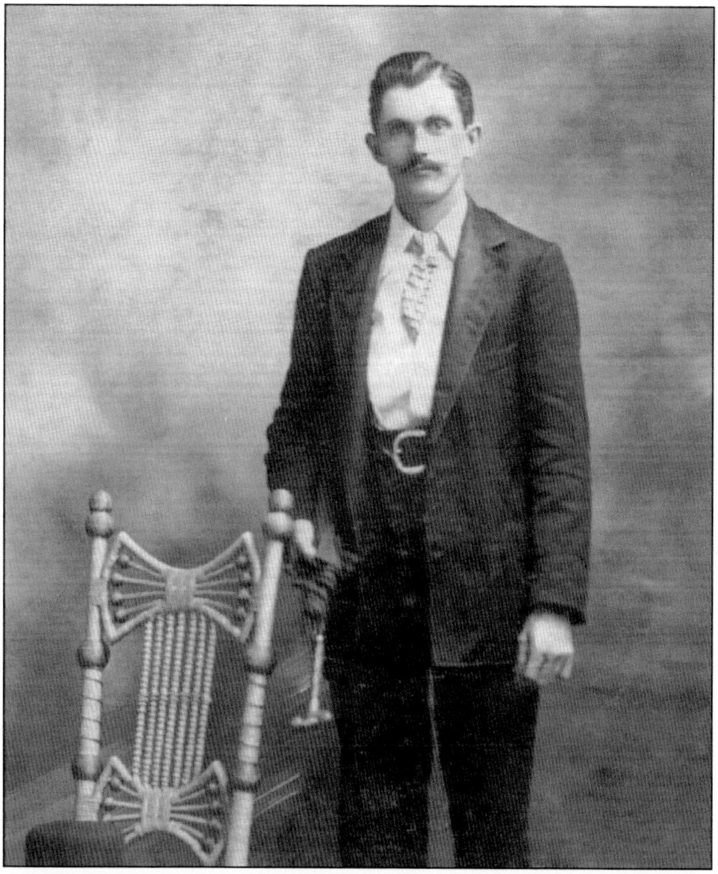

H.H. Goode helped his son hide his distilling equipment in one abandoned mine. When he heard revenuers coming, he walked over one mile through the twists and turns of the mine without light to warn his son of the intruders. Family members said they would have starved if not for the selling of liquor to residents of Birmingham. (Author's collection.)

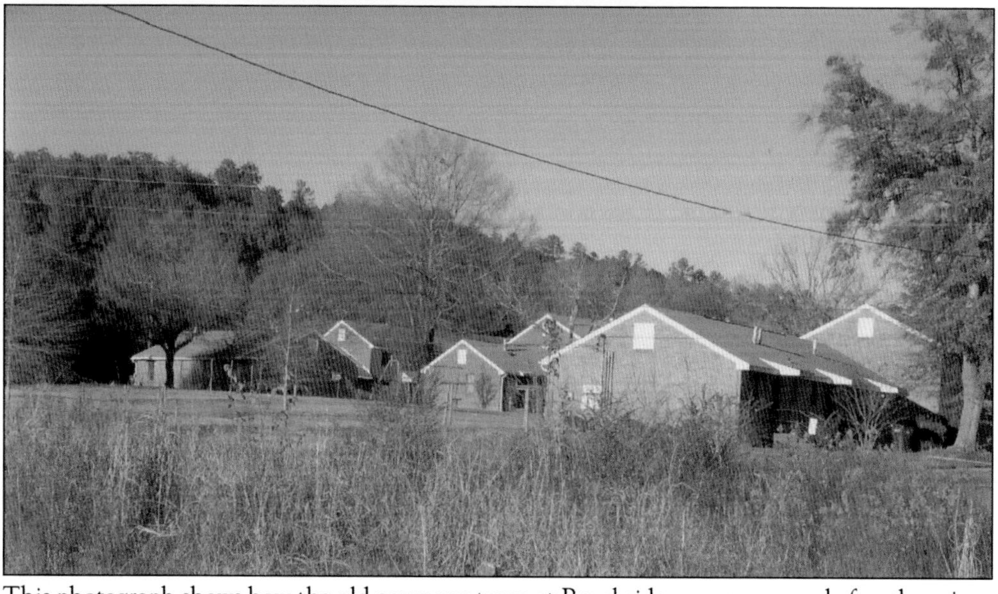

This photograph shows how the old company town at Brookside was repurposed after the mines closed. In the background, government apartments stand where company buildings and houses once did. (BHM.)

Unlike life in the city, coal towns rarely had running water early on. Brookside was one exception that had its own water system. Most homes had chamber pots and outhouses like this one. (LOC, FSA.)

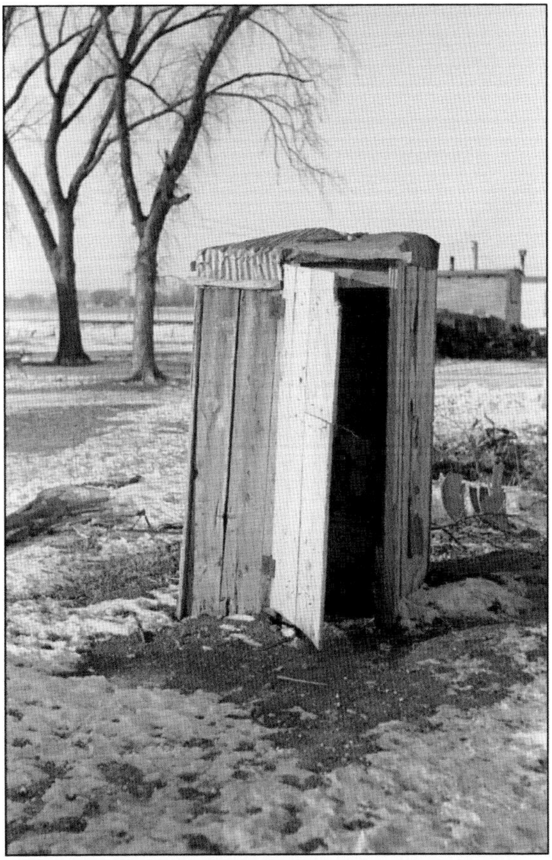

This modern house was more indicative of an urban company dwelling. Its styling was practical in design and its finished underpinning indicative of a more polished approach to company living. (LOC, HABS.)

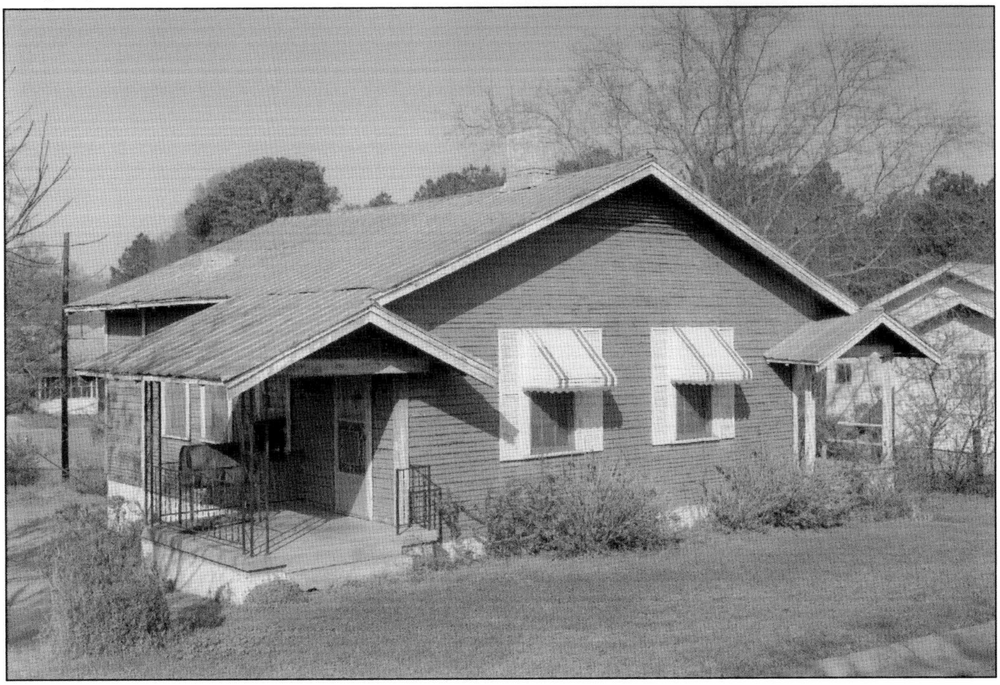

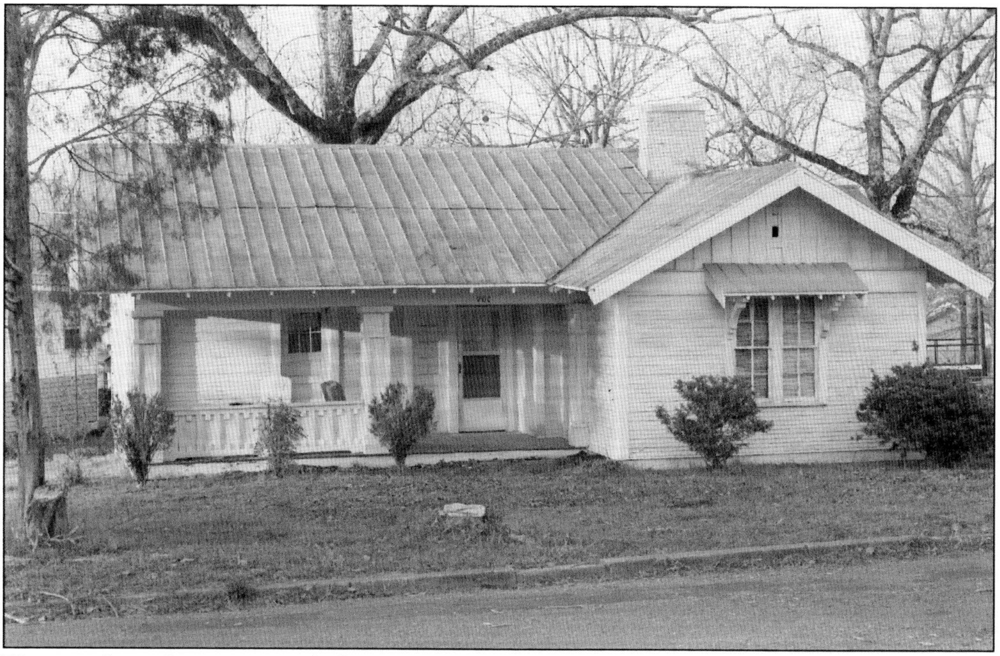

This nice bungalow was typical of TCI's corporate welfare approach to home building. Nice homes were an incentive to keep workers happy and loyal to the company. (LOC, HABS.)

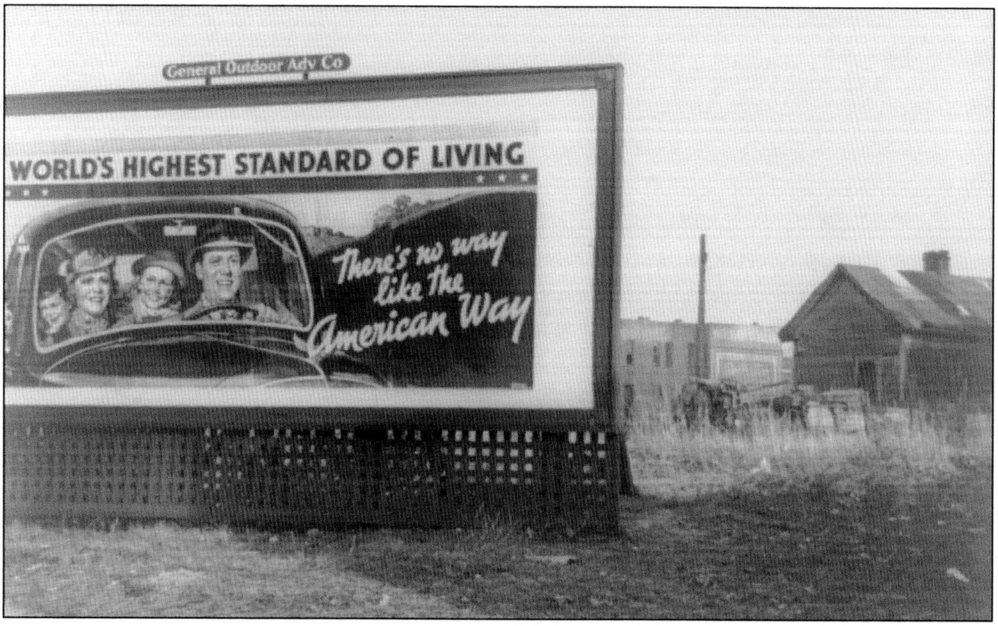

The Resettlement Administration operated from 1936 until 1937 when it was repurposed into the Farm Security Administration. It moved hundreds of families from rural and urban environments to planned cities. Notice this sign hailing a high standard of living with shacks in the background. Franklin Roosevelt's New Deal helped to make this dream possible for many. (LOC, FSA.)

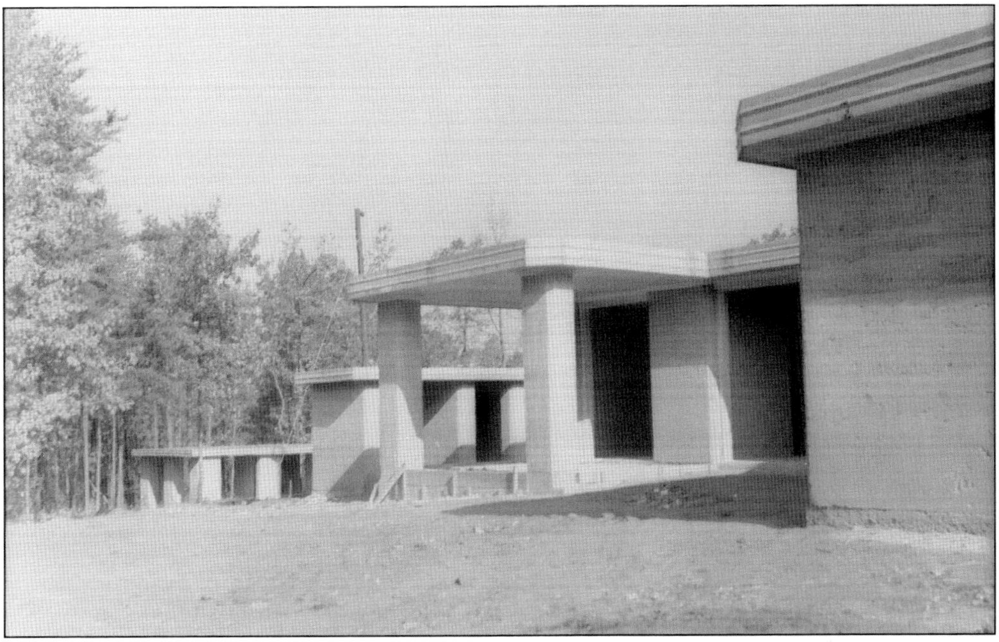

The Resettlement Administration (RA) bought rural and urban homes that were considered poor or in blighted areas. They then resettled these families in model communities like Gardendale. This is one example of the new homes. (LOC, FSA.)

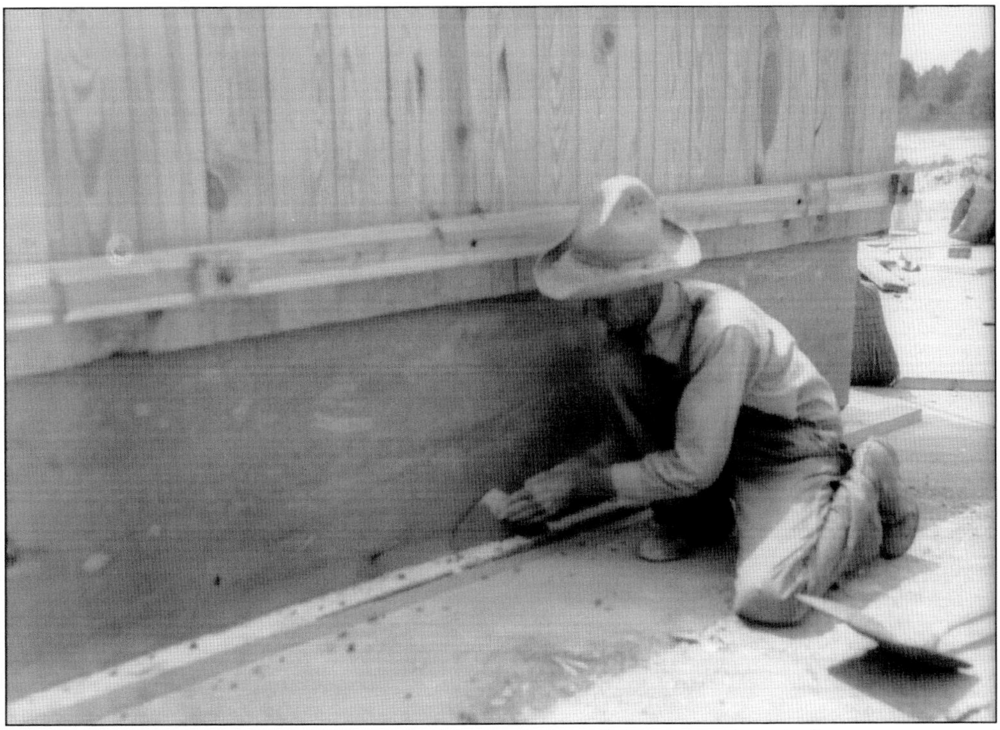

One portion of the RA project was to build rammed earth houses and settle former miners and others into them. Construction on this rammed earth house in Mount Olive was a cooperative effort between the family moving in and the contractors. (LOC, FSA.)

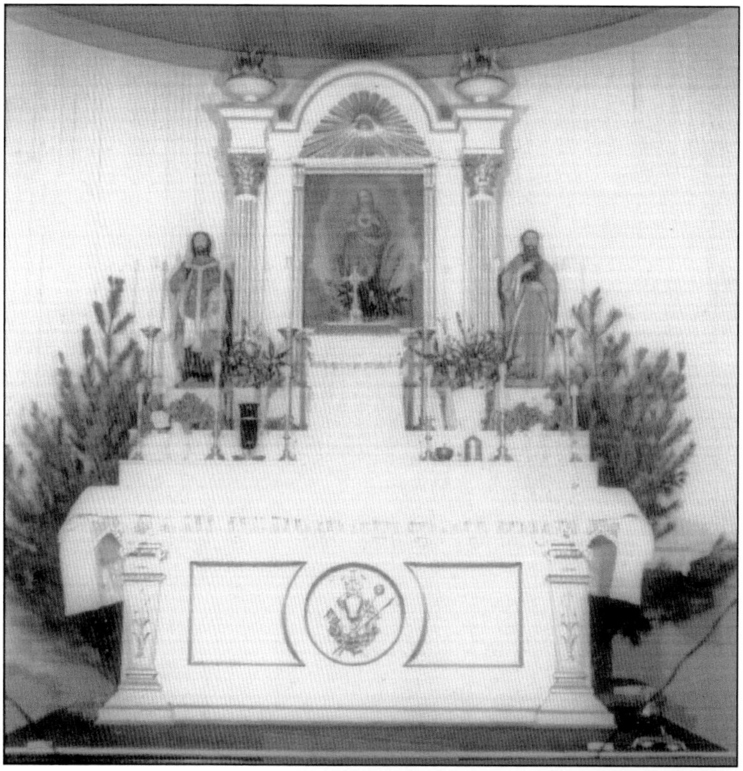

One of the ties that bound immigrant miners together was religion and the establishment of practices familiar to them. From the earliest of days, immigrant miners across the district built Roman Catholic churches and, in Brookside, a Russian Orthodox one. For immigrant mining families, religious tradition helped them blend old and new. (BHM, Albert Billits.)

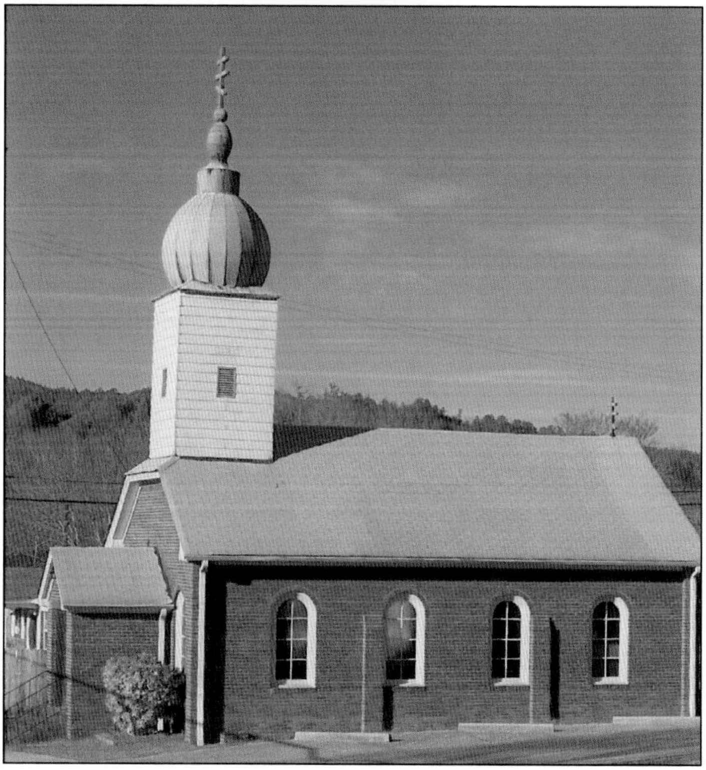

Slovak immigrants brought with them to Brookside a unique church never before seen in the South. They founded the first Russian Orthodox church south of the Mason-Dixon line. Organized in 1894, the Slovak church at Brookside served many Slovak families in the area, and until the 1960s, mass was conducted in the original language. (Author's collection.)

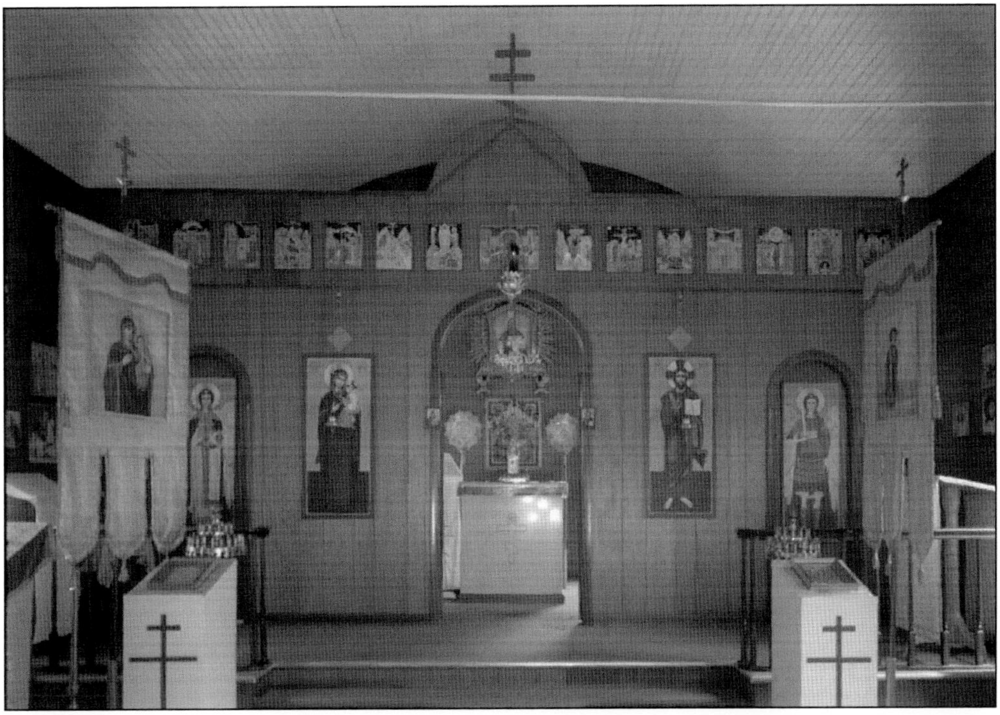

St. Nicholas is a unique sight in the Alabama countryside. Its cupola dome dominates Brookside's skyline and its interior is a menagerie of precious icons. (St. Nicholas Russian Orthodox Church.)

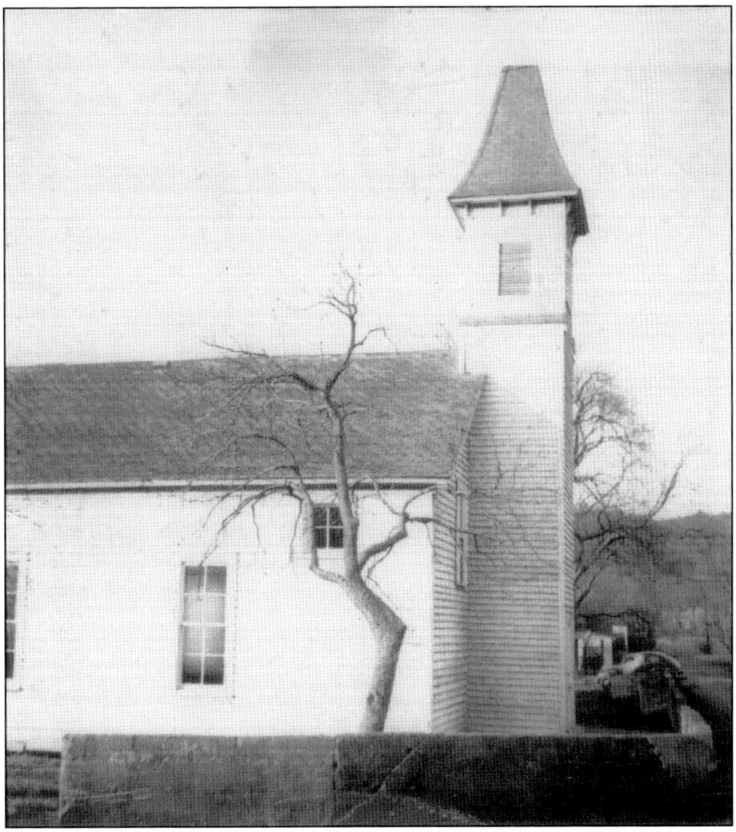

Religious life was vital to all in the mining communities. This is SS Cyril and Methodius Catholic Church, built in 1894 by immigrant miners and their families. (BHM, Albert Billits.)

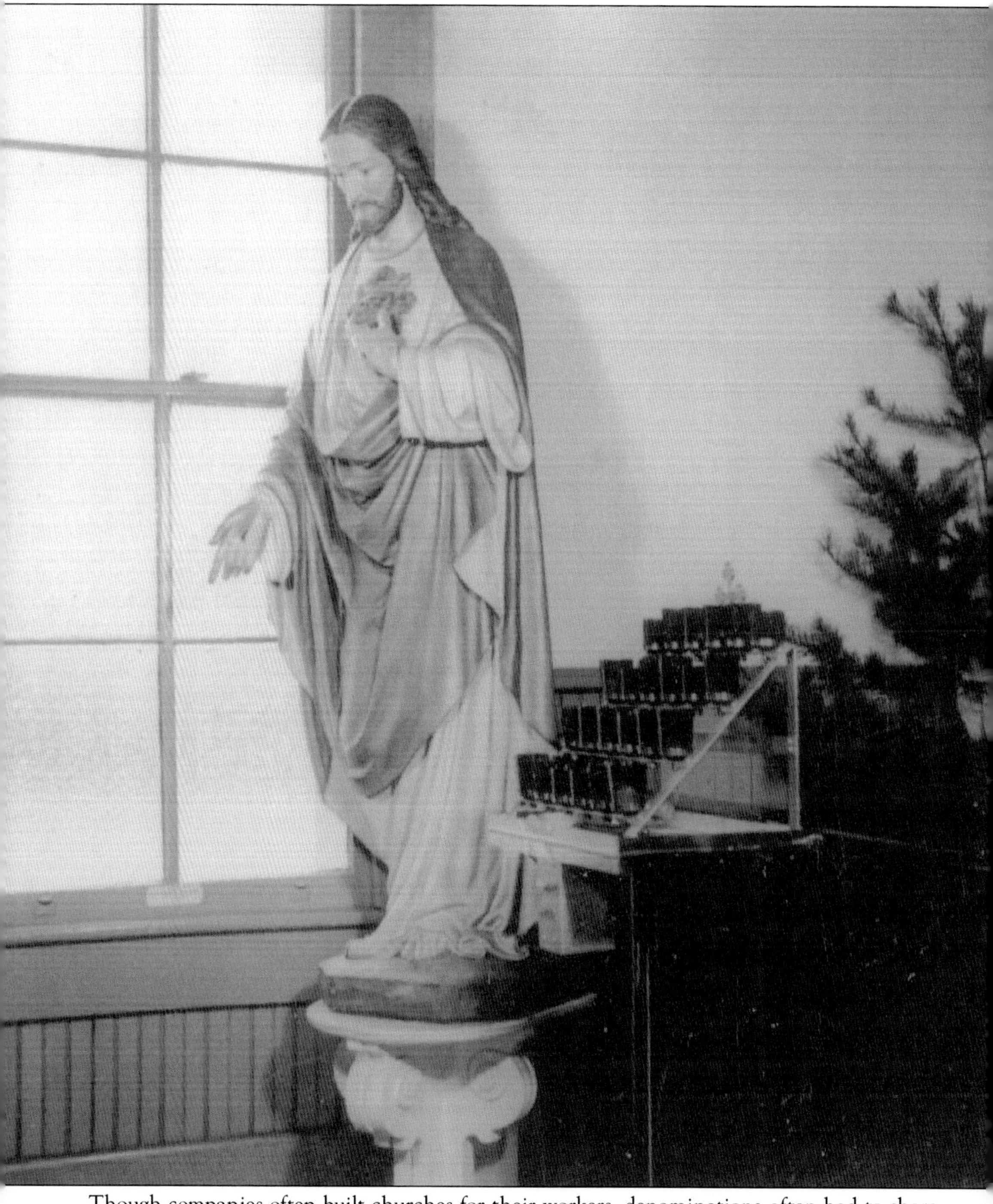

Though companies often built churches for their workers, denominations often had to share one facility, alternating Sundays. This alter piece is from a privately built church in Brookside. (BHM, Albert Billits.)

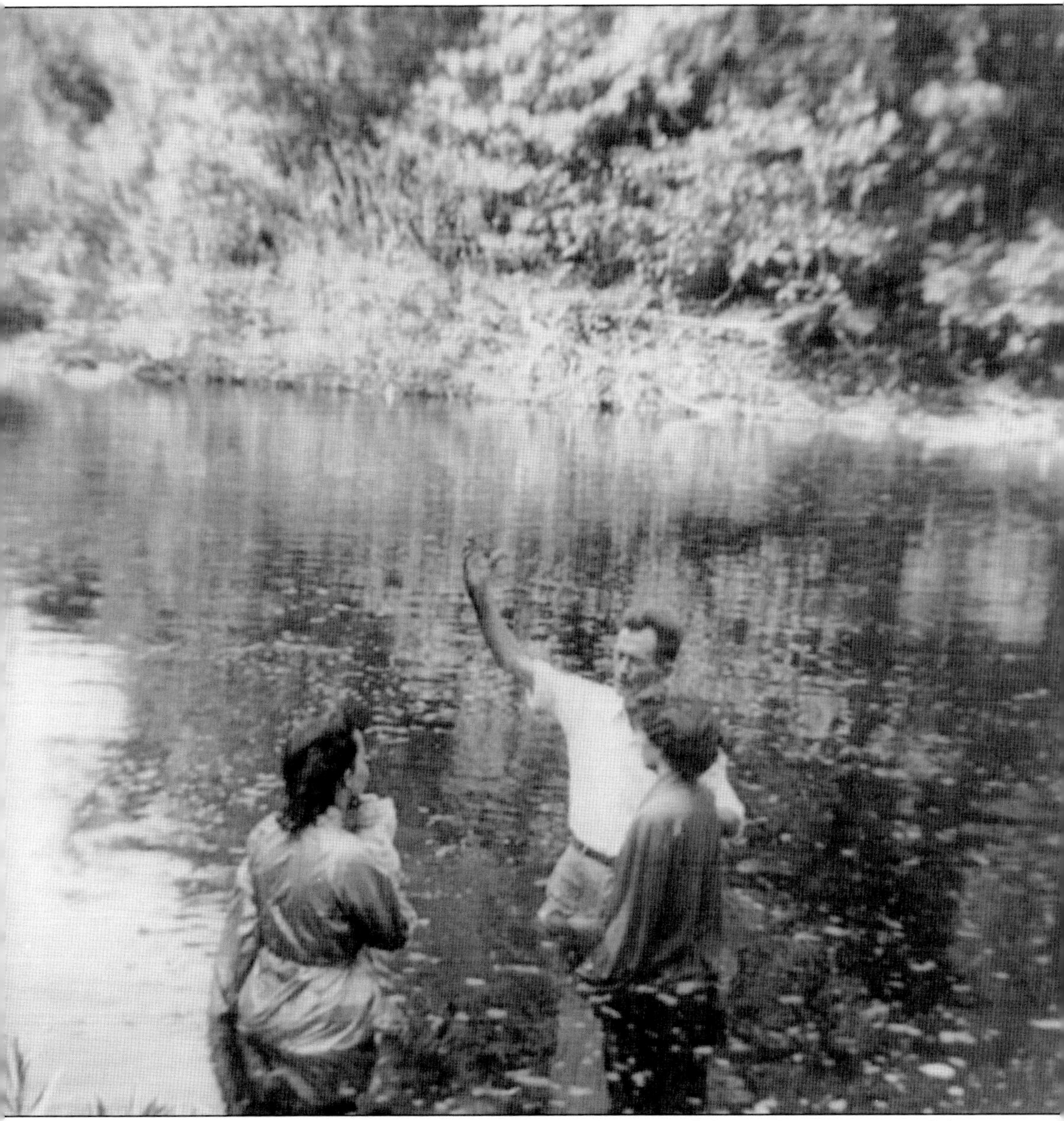

Five Mile Creek, actually much longer than its name implies, served the coal communities of the county well. Not only did the water service farmers' irrigation systems and coal operators for fire suppression, it served churches as a baptismal pool. Each community had its own special place for baptisms, usually done en masse. (Ralph Sanford.)

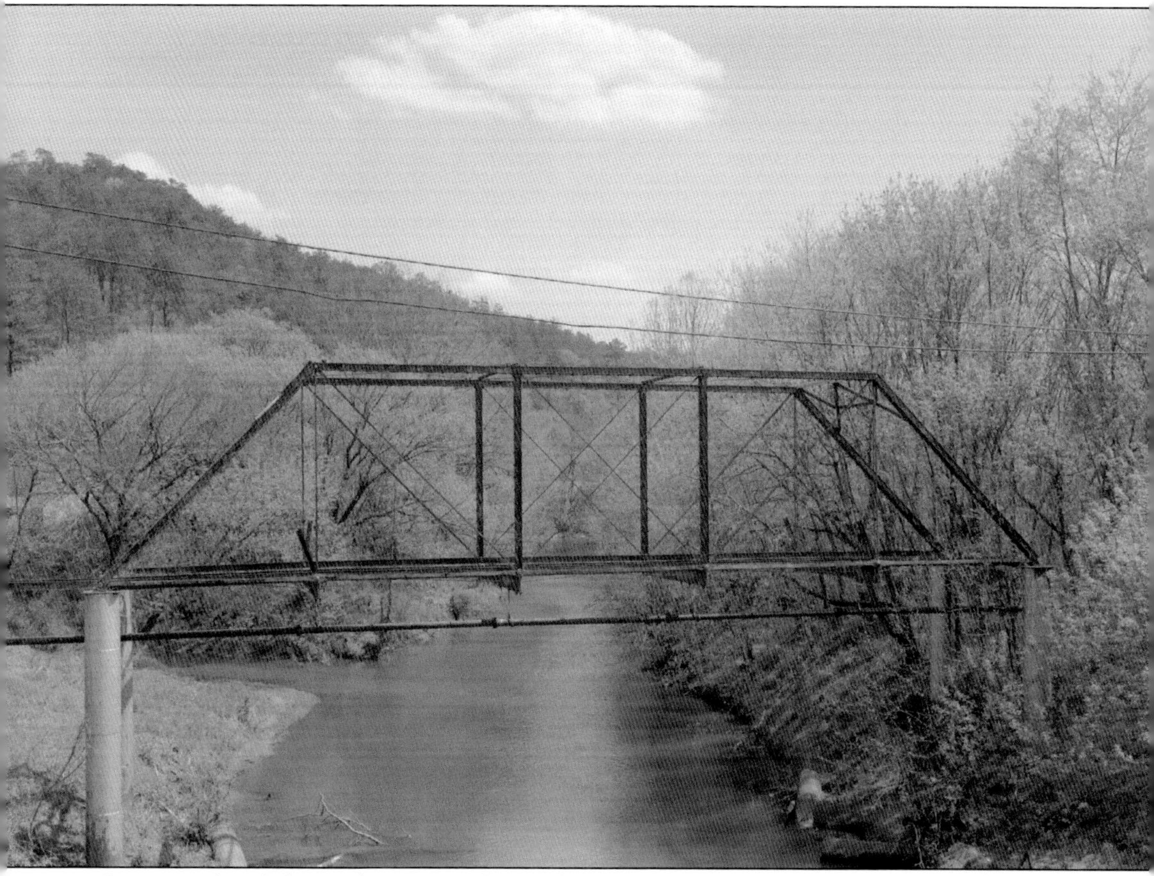

Interestingly, while people went to the creek to wash away their sins, the creek was known as Creosote Creek because of the industrial discharge that was dumped into it. People reported coming out of the creek with an oily sheen. (Author's collection.)

TCI built recreational, educational, and religious facilities for its employees. It also built company churches for both its white and black workers. This church was for its white workers in the Mulga area. (LOC, HABS.)

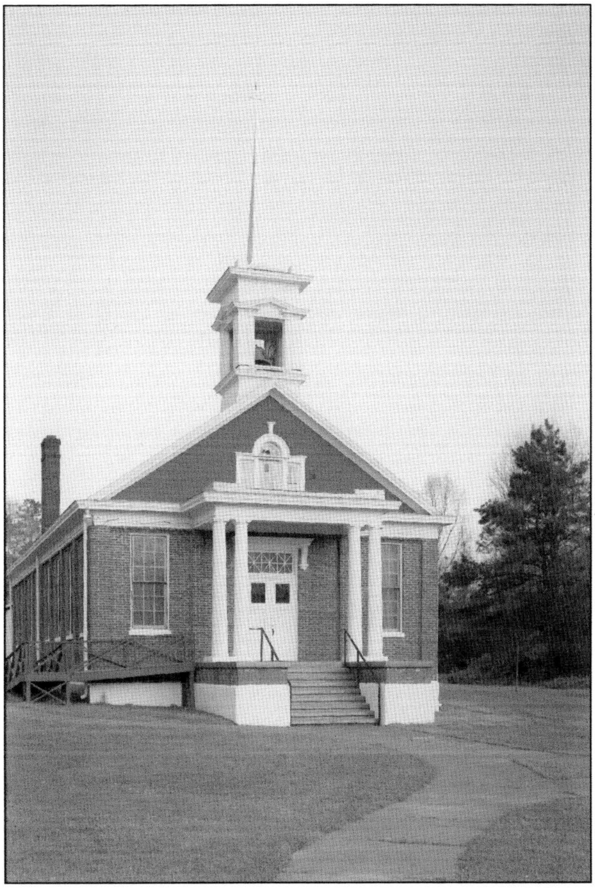

Notice that the TCI church for blacks near Bessemer was built with wood siding, not brick, and was not as large and imposing as the white church. (LOC, HABS.)

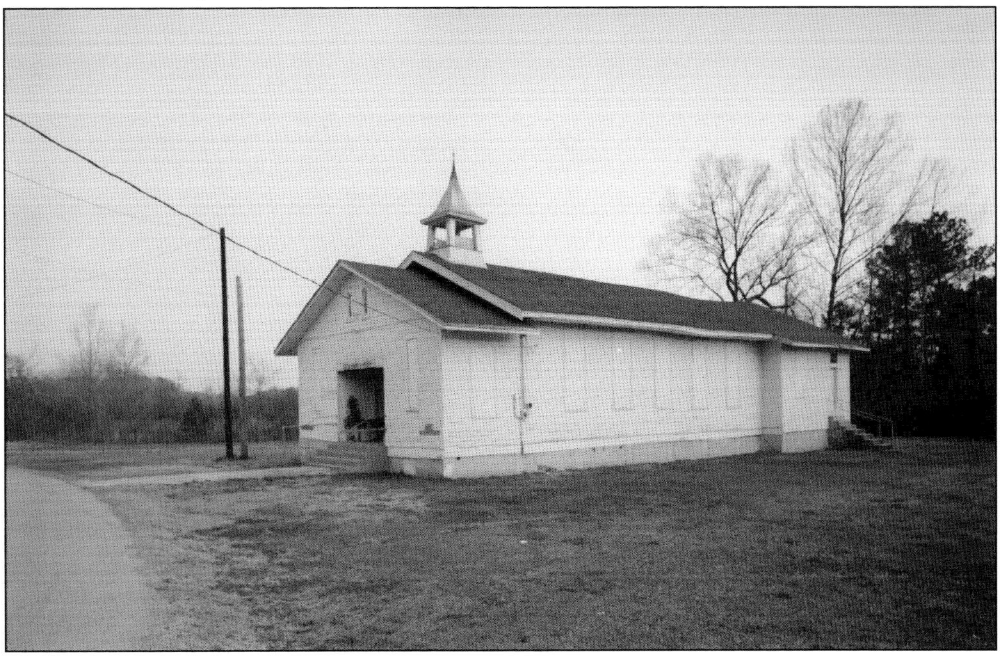

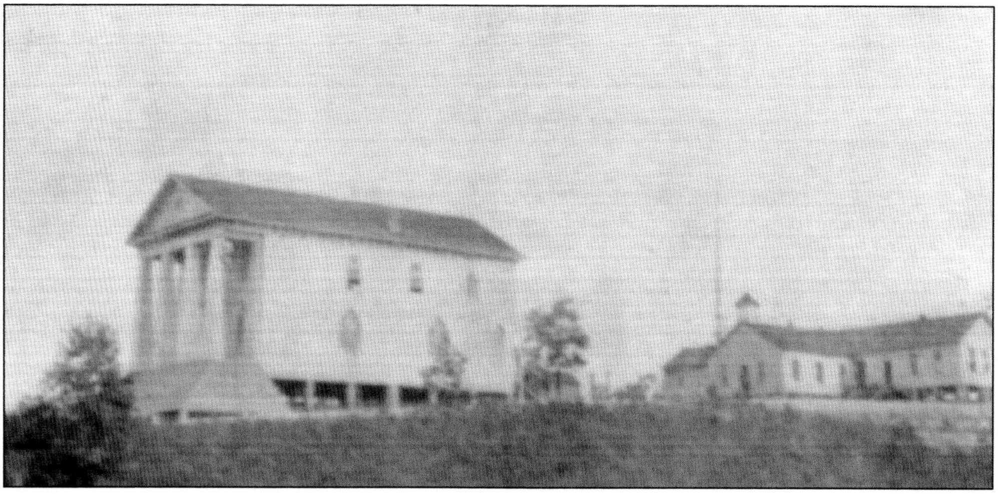

Fraternal organizations were extremely important among the county's coal miners. Amid poor labor-management relation guilds and fraternal organizations such as the Knights of Pythias, the Masons and other groups gave miners a place to come together to air out their troubles and fraternize. This guild hall at Bessie was quite impressive. (Author's collection.)

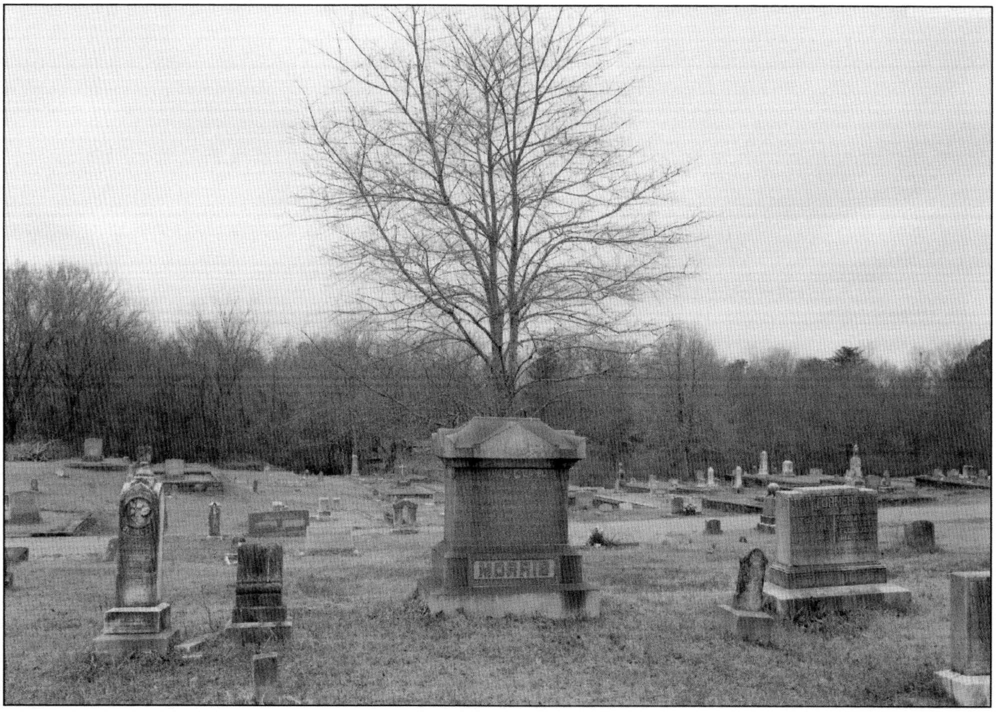

Each community had its own division based on places of origin. Blossburg had Cherokee Row because of the large numbers of people from Cherokee County who migrated there. Pratt City had Irish Hill and Catholic Hill. These divisions even spread over in final resting places, as Fraternal Cemetery was the final resting place for hundreds of immigrants. (LOC, HABS.)

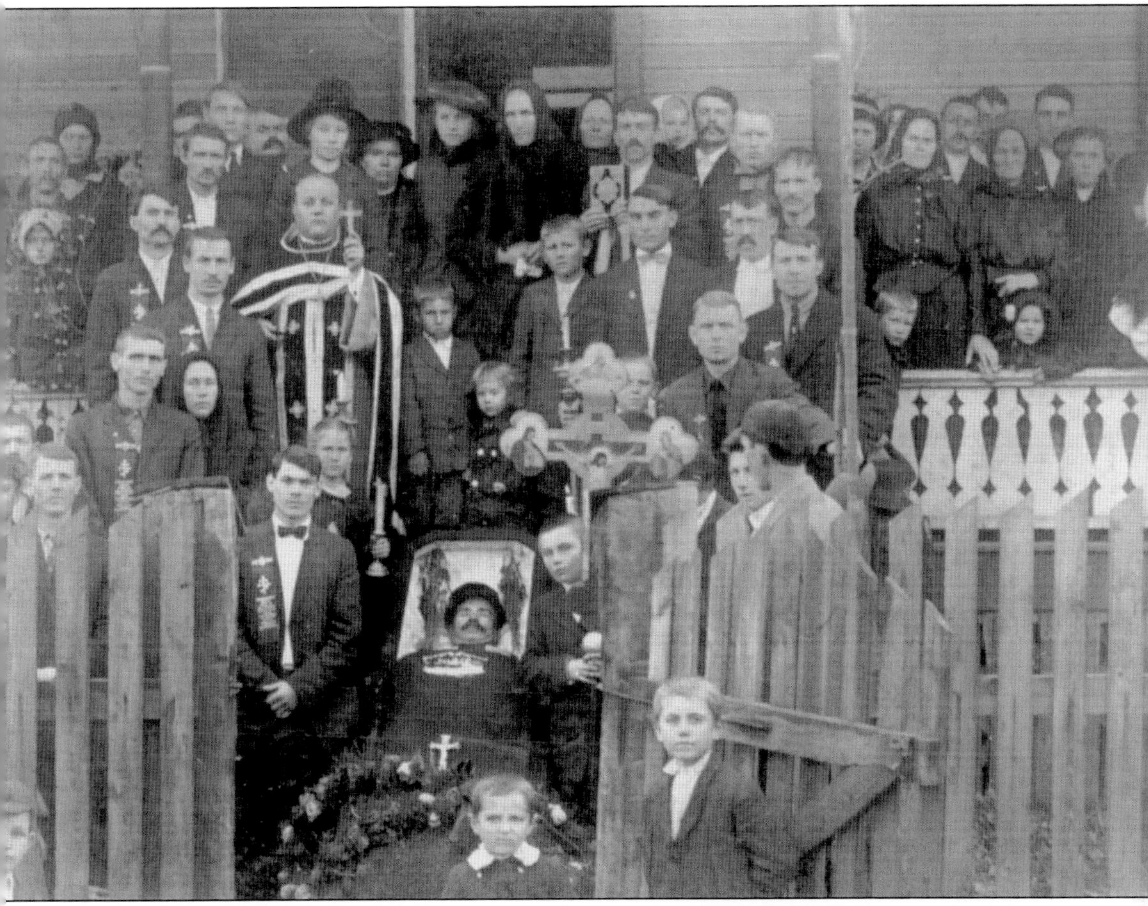

Slovak funerals in Brookside affected not only the Slovak community, but also the schedule of the entire community. Funerals lasted three days, and so many of the miners in Brookside were Slovak that the mines had to close on those days, as there was not a sufficient workforce to operate them. (BPL Collection.)

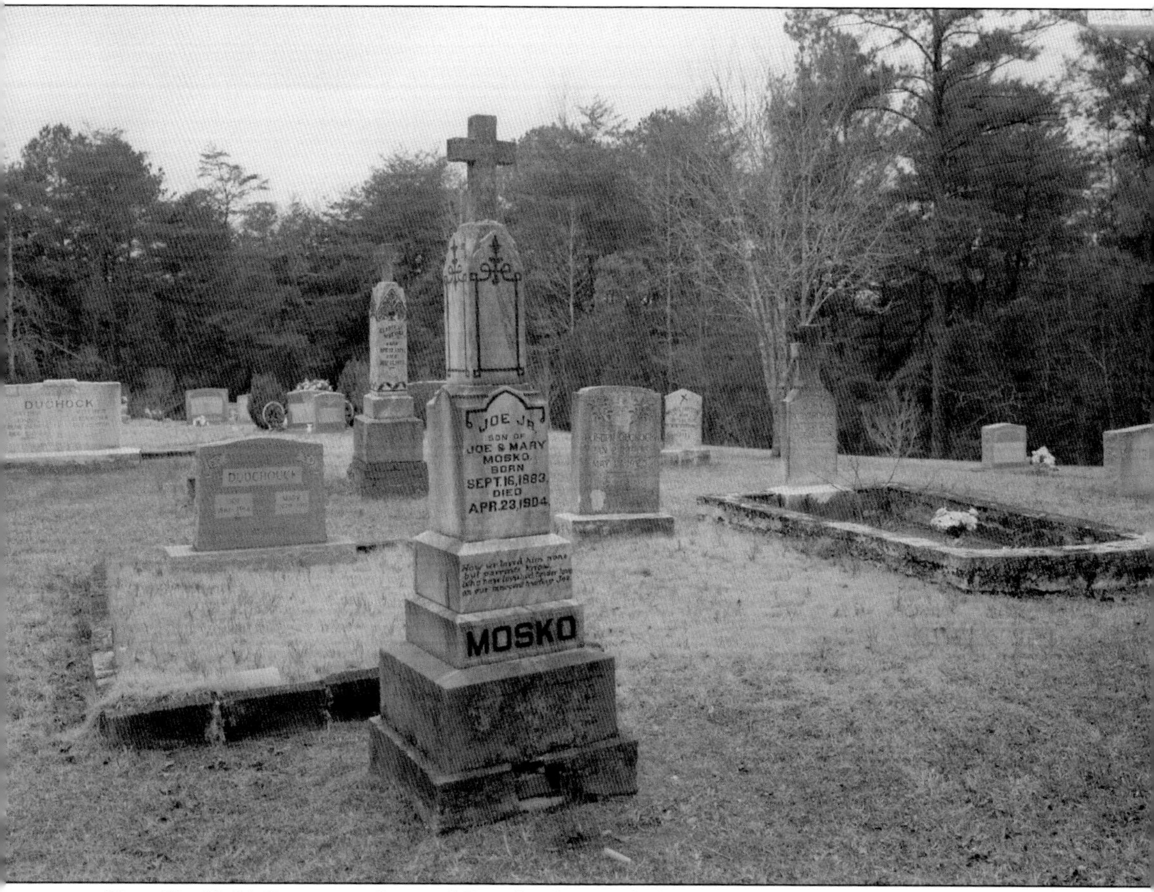

Sloss-Sheffield Company donated Brookside's Catholic cemetery for use by its employees and their families. Michael Ziach, an immigrant who worked in the mines, petitioned the company for the land and, ironically, was the first to be buried there when he died of typhus in 1901. (LOC, HABS.)

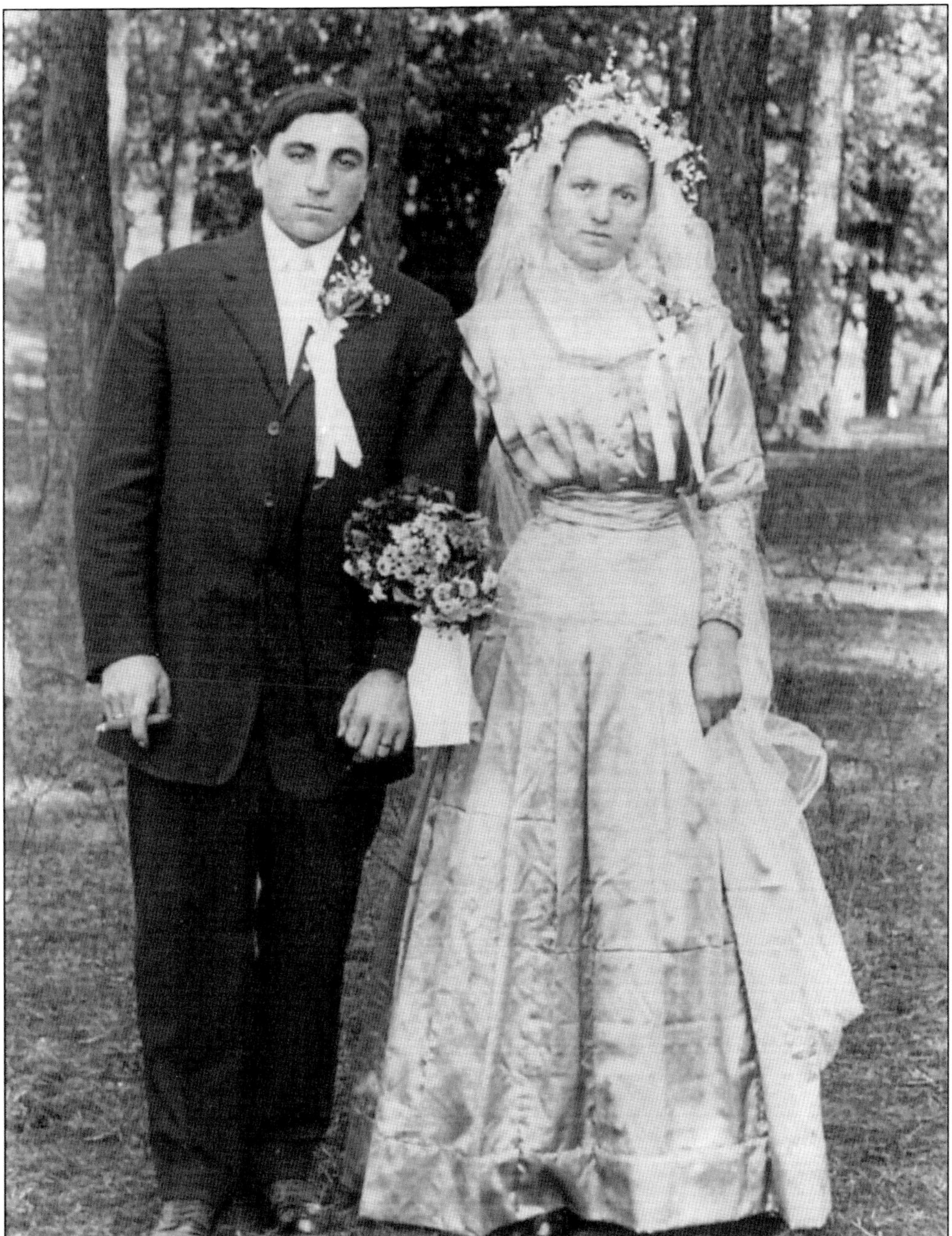

Immigrant miners' weddings were an important means of celebration, and these events also affected mine work schedules. Accustomed to three-day celebrations, beer and wine flowed freely, and since the majority of the community attended, the mines were forced to close temporarily. (BHM.)

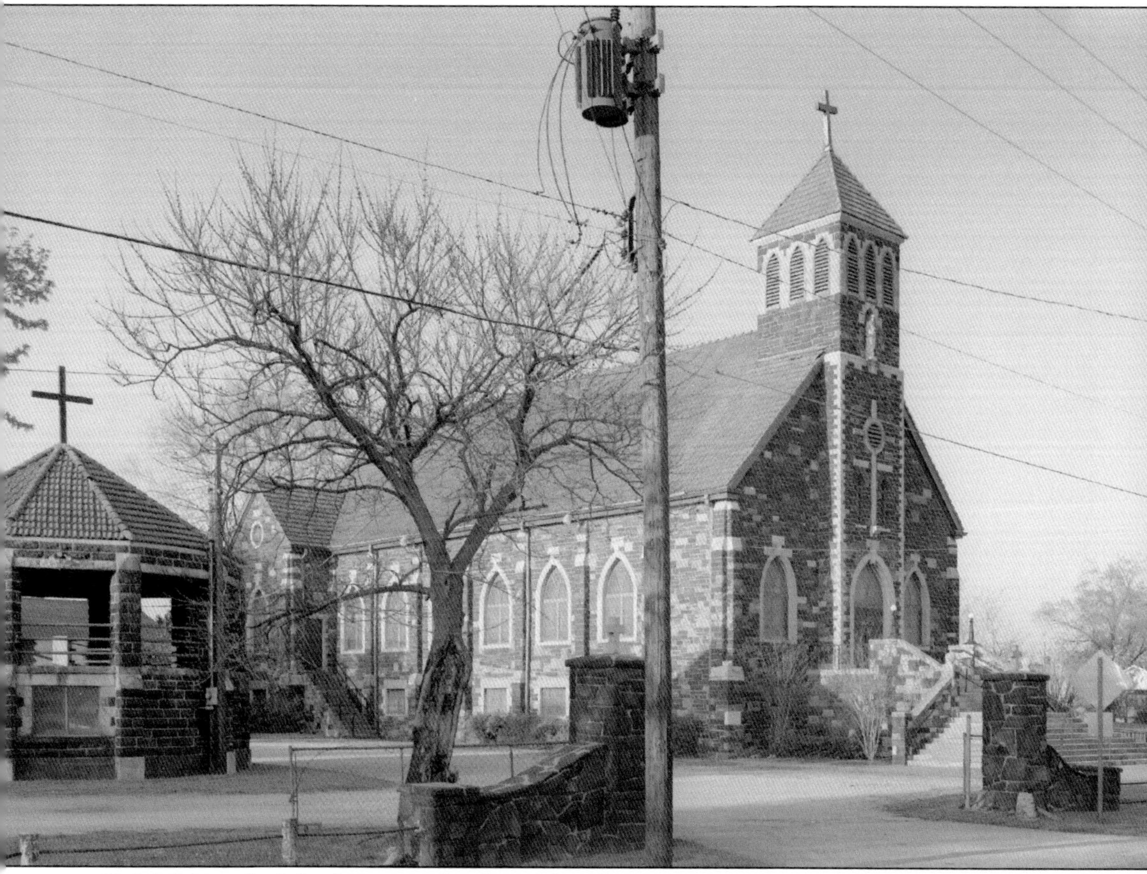

St. Mark's was built by the mining families of the Thomas mining area. It was built of locally quarried rock. St. Mark's served Italian immigrant families until its closure in the 1970s. (LOC, HABS.)

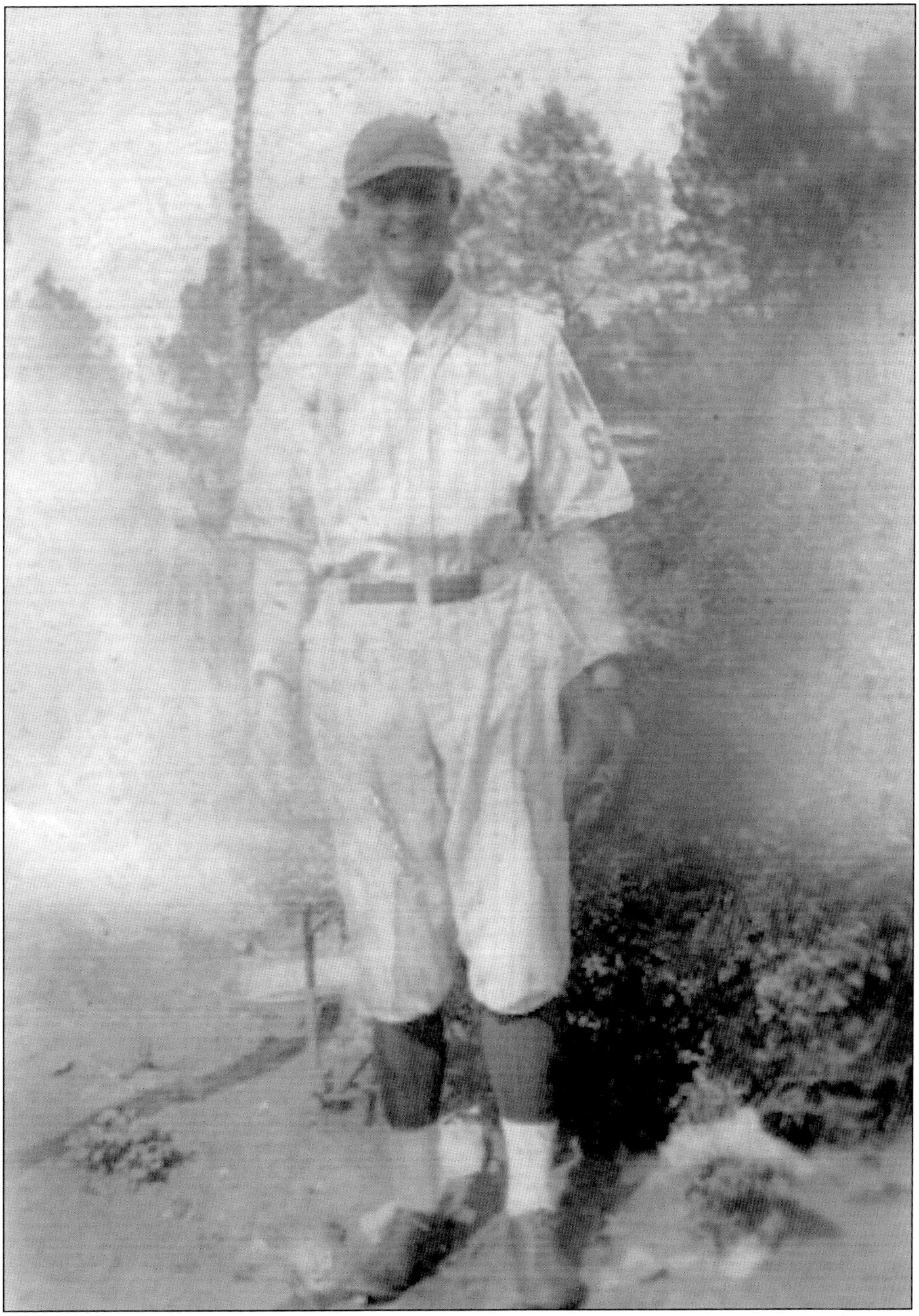

John Clifton "Pud" Goode was one of many young men who played baseball on a semiprofessional basis in and around Jefferson County's coal towns. So fierce was the competition among these industrial league teams that many brought players in to play. Labuco brought Goode in not to mine coal, but to play left field for their baseball club. (Author's collection.)

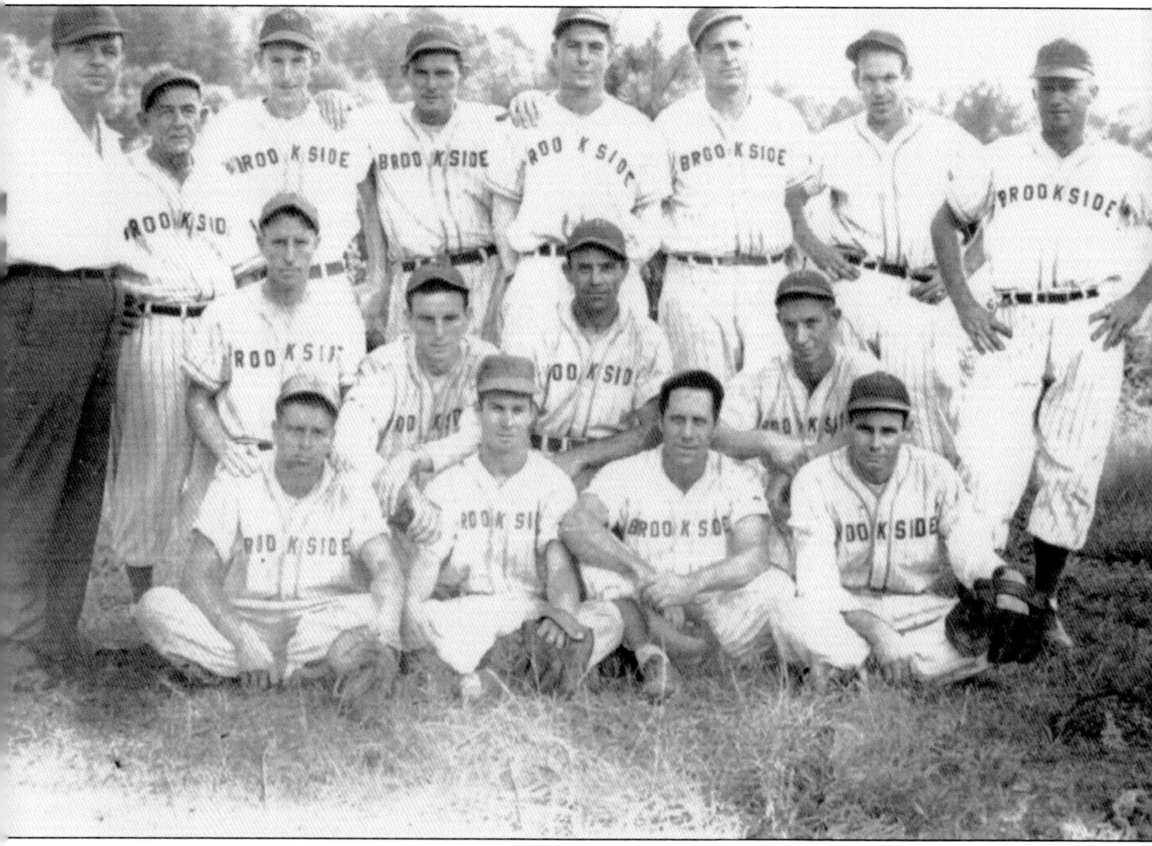

Miners needed a release from the pressure cooker of mine work, and many found baseball to be the answer. Often men left work and organized a pickup game on the way home. On Sunday afternoons, the crack of the bat attracted many to the baseball diamonds all across the district. (BHM, Albert Billits.)

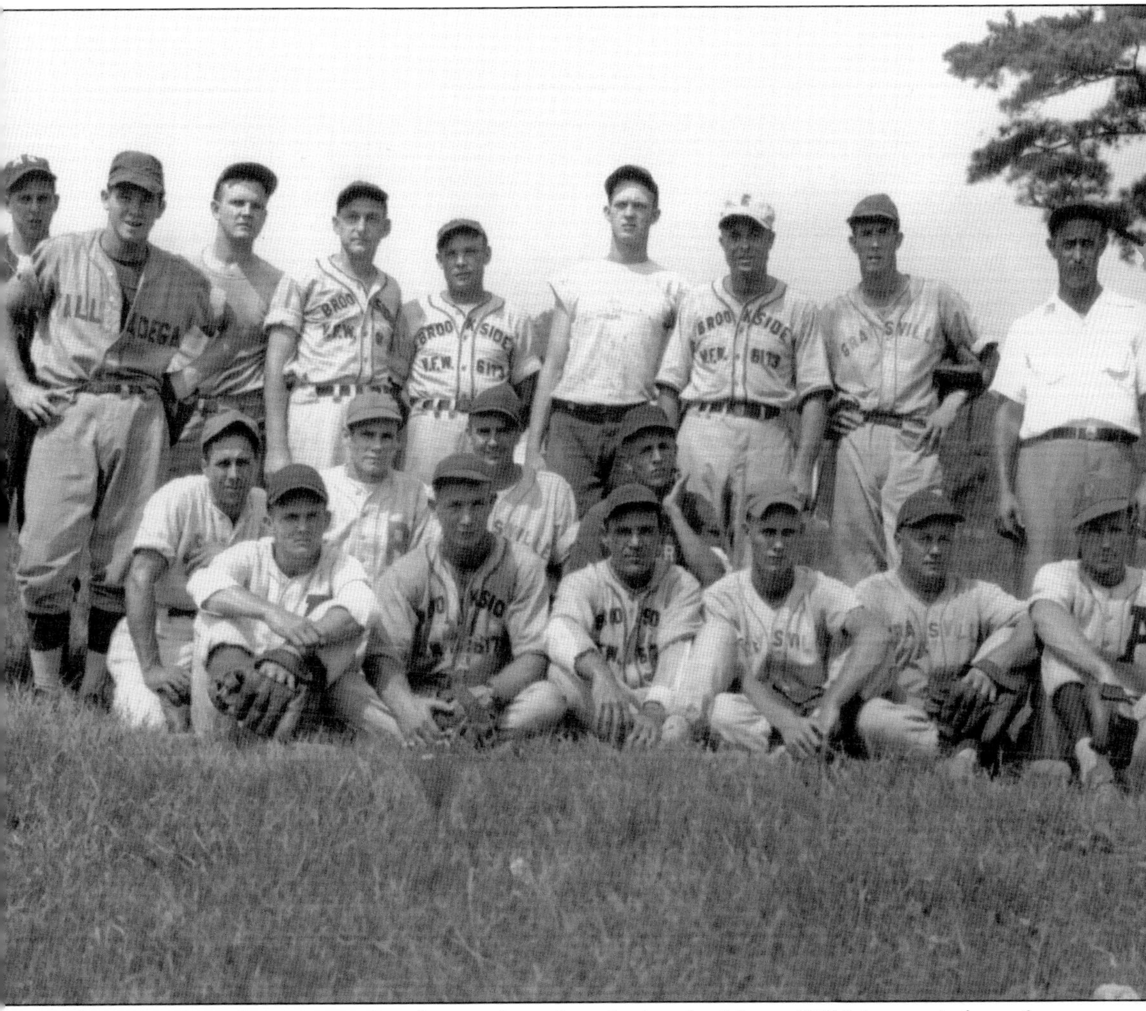

The county's coal towns produced a number of professional athletes. Bill Moore tried out for Milwaukee, and his brother Charles "Mule" Moore tried out as catcher for another professional team. Charley "Butch" Moore played major league baseball for 20 years. (BHM, Albert Billits.)

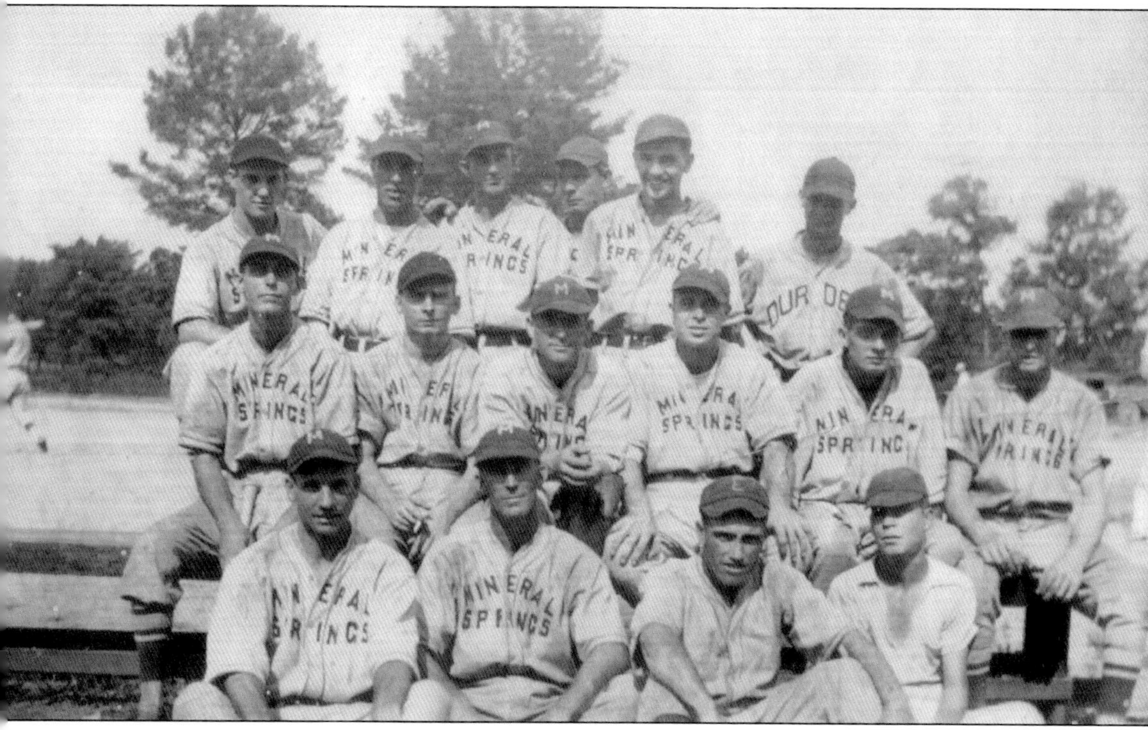

Each community had a baseball diamond, and this Mineral Springs team dominated theirs. They were pennant winners in their league. (Author's collection.)

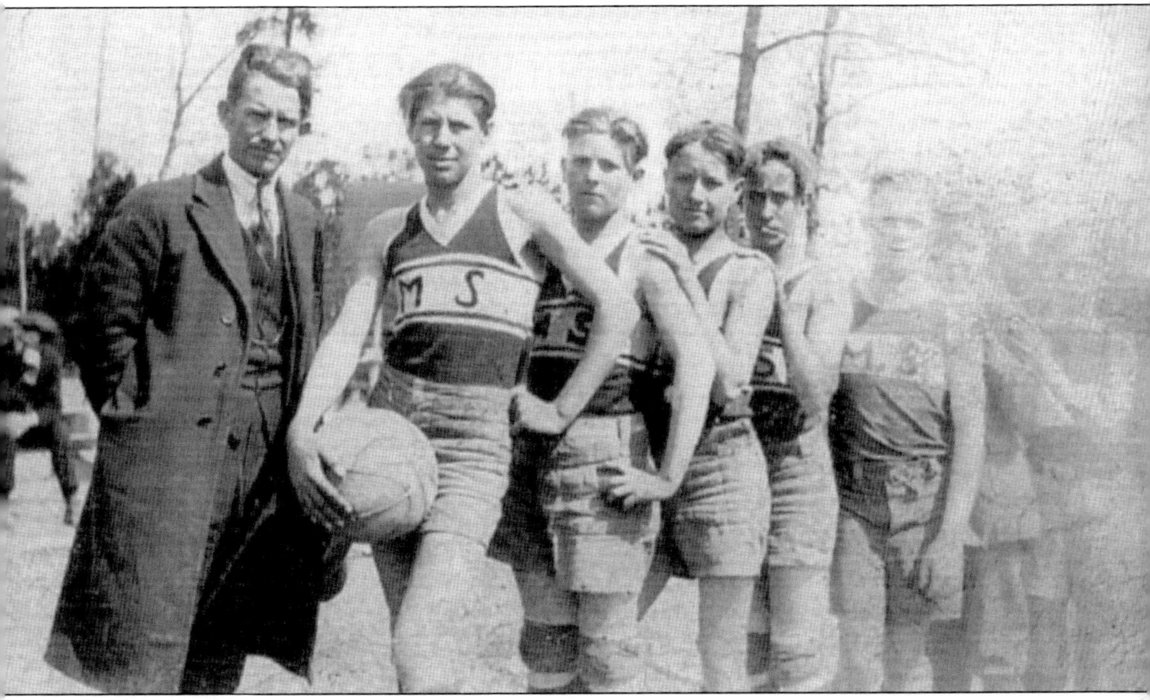

Many school-age boys also enjoyed the game of basketball. The mining towns of Mineral Springs and Brookside put out several state championship teams. (BHM.)

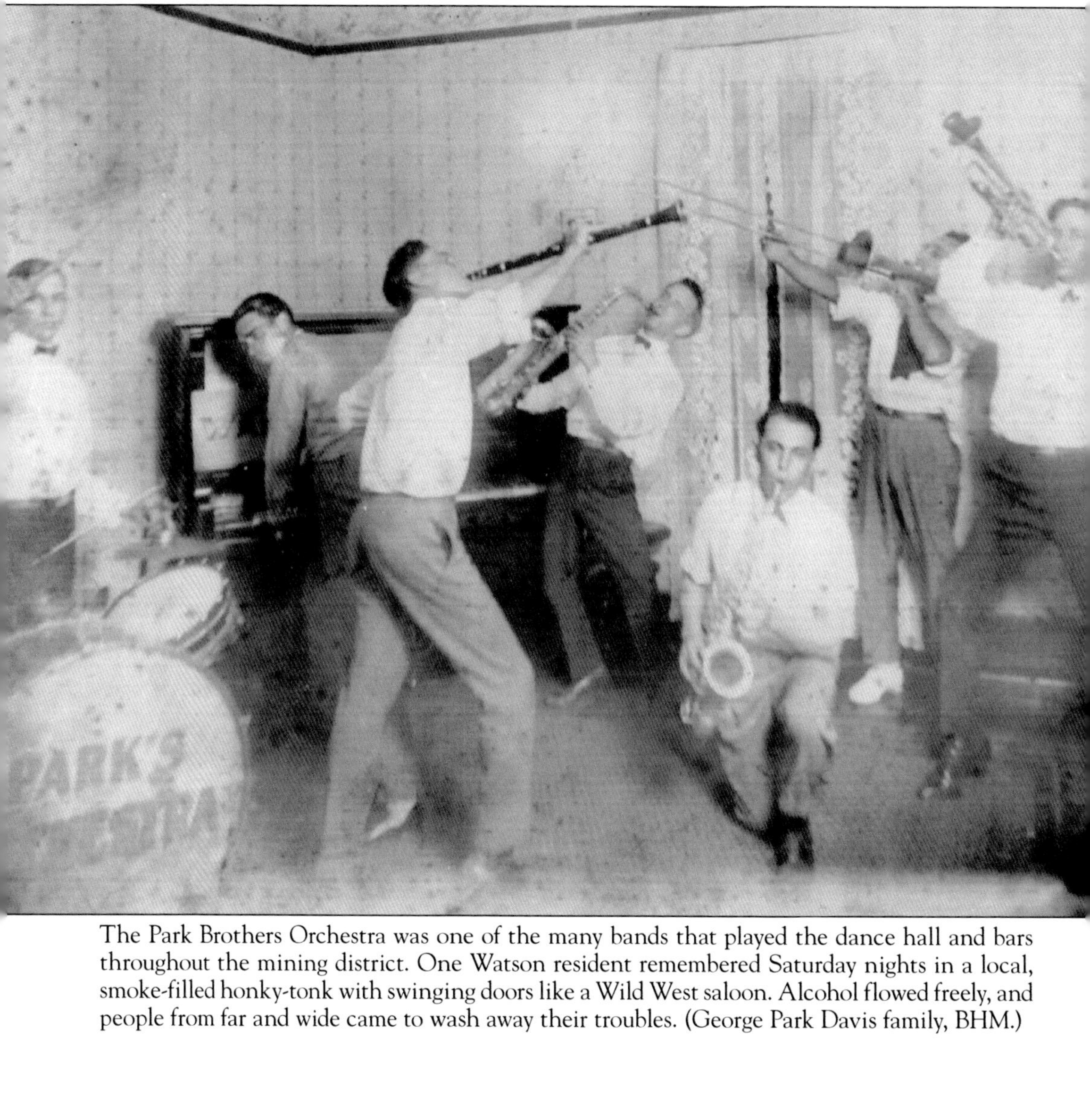

The Park Brothers Orchestra was one of the many bands that played the dance hall and bars throughout the mining district. One Watson resident remembered Saturday nights in a local, smoke-filled honky-tonk with swinging doors like a Wild West saloon. Alcohol flowed freely, and people from far and wide came to wash away their troubles. (George Park Davis family, BHM.)

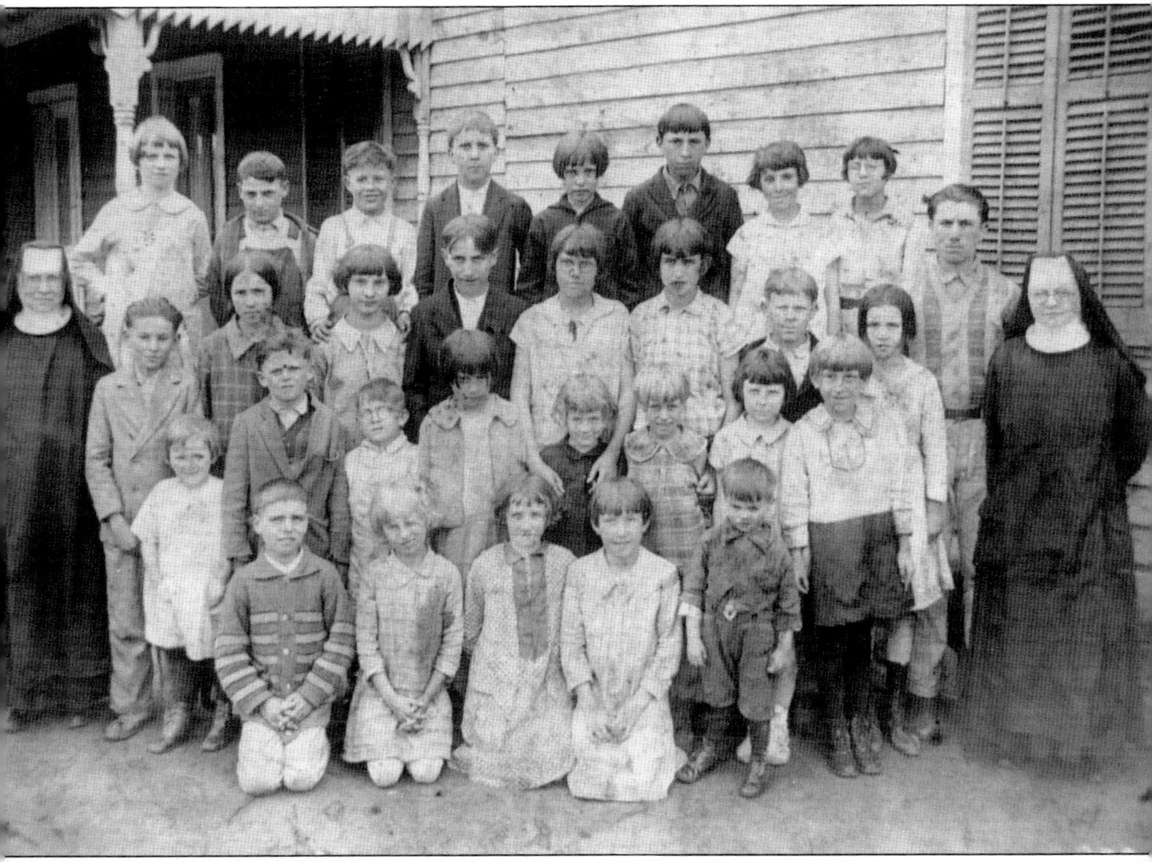

While Brookside had a public school, the immigrant miner families established two parochial schools. One was St. Joseph, a Roman Catholic parochial school that closed in 1935 because of dwindling enrollment. (BPL Collection.)

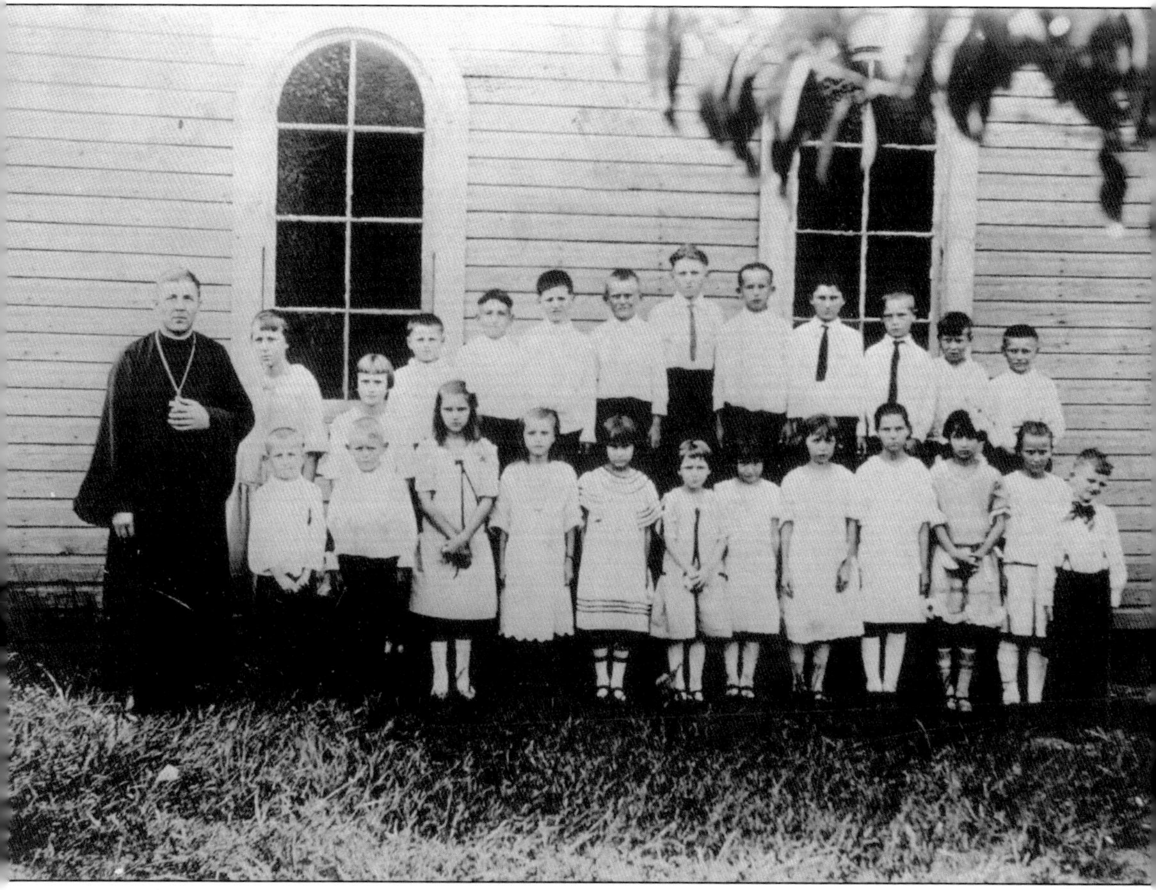

Brookside's other parochial was the St. Nicholas Russian Orthodox School, pictured here. It too closed among dwindling student numbers. When immigrant students were thrust into the public school system, they experienced culture shock. Many of their classmates teased their style of dress, skin tone, and language. (BPL Collection.)

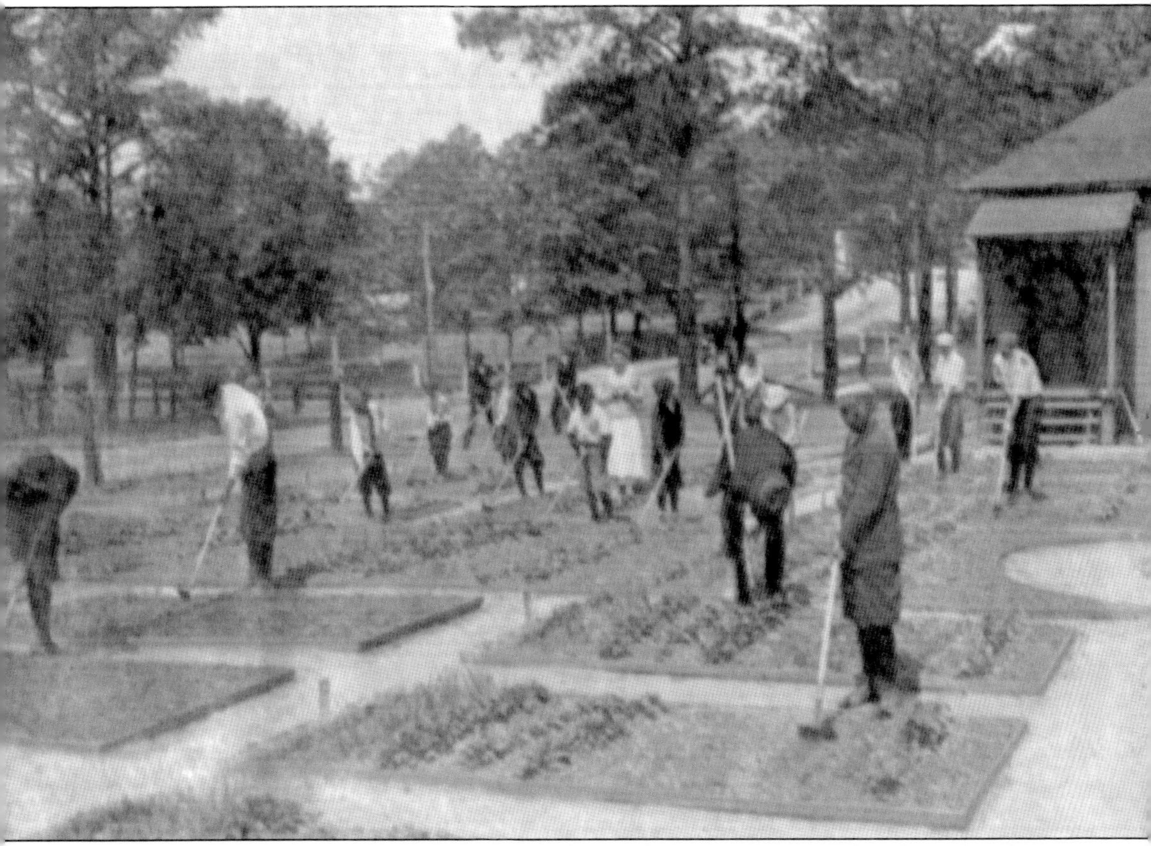

TCI planned a series of model communities throughout Jefferson County. TCI hired social workers and teachers and built community centers, schools, and recreational centers to enrich the lives of workers and their families. Here, teachers are promoting an exercise in planting. Most families had home gardens to supplement their food supplies. (Author's collection.)

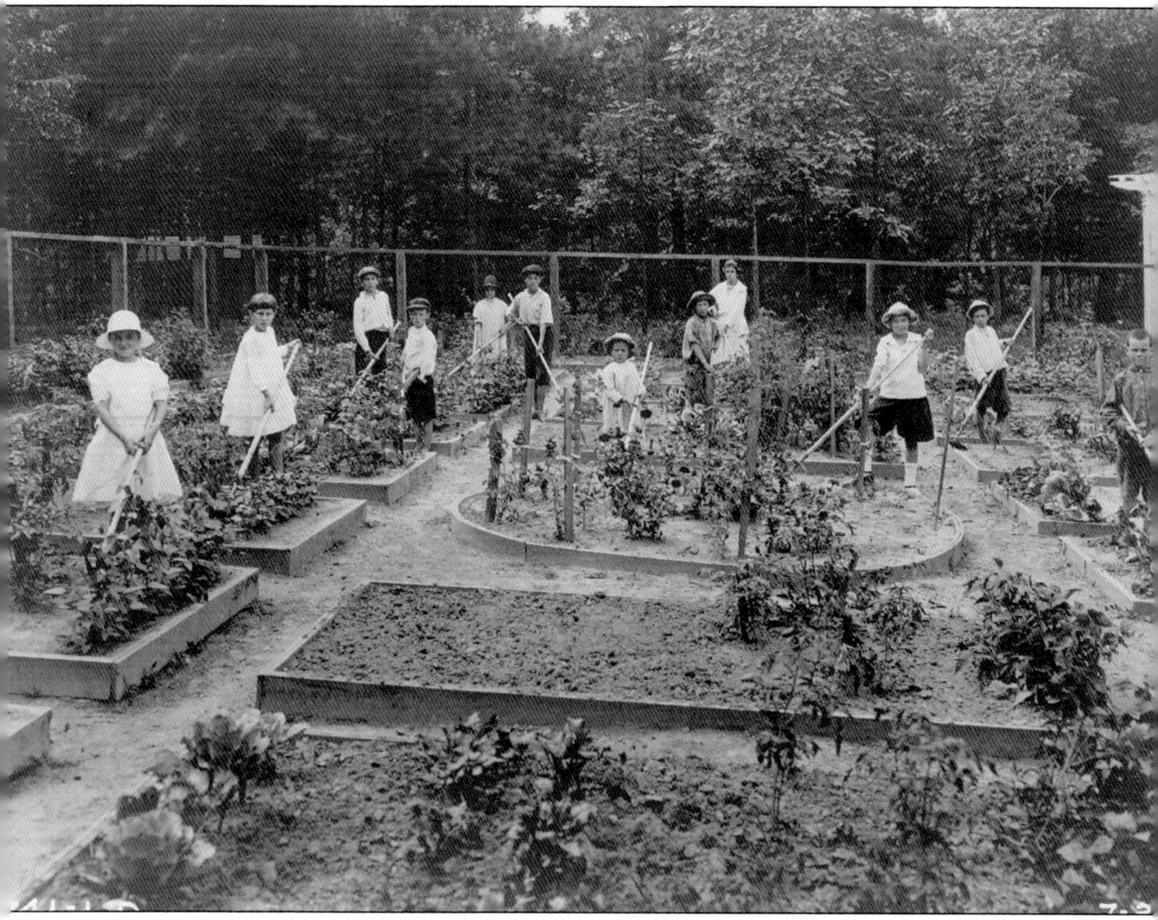

This early photograph of the horticulture class at TCI's Edgewater School was taken during the last push of America's involvement in the First World War. The cultivation of Liberty Gardens was deemed a patriotic duty. (LOC.)

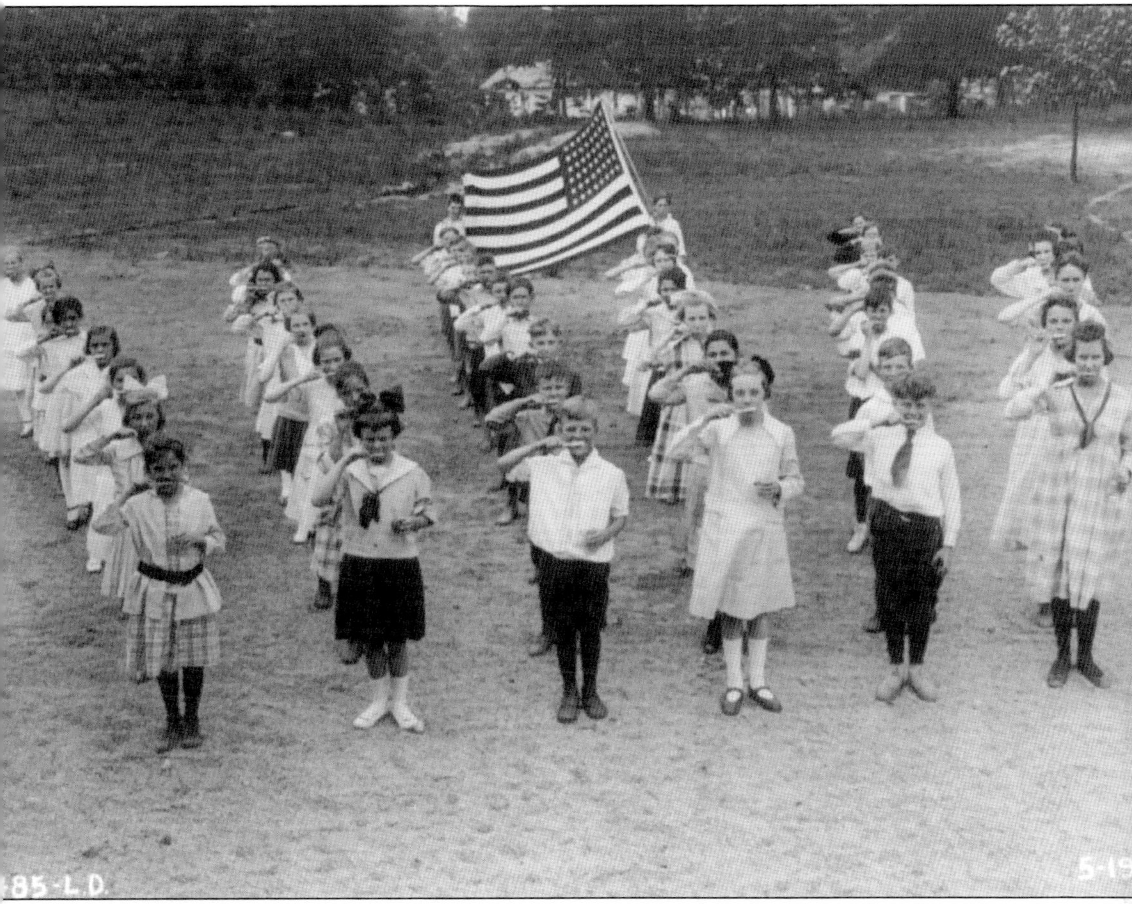

TCI, known for welfare capitalism, attempted to better its employees and their families in every facet, including dental health. Here, a schoolteacher leads a group of children in a toothbrush drill at one of the company schools. The flag in the background suggests health is akin to patriotism. (LOC.)

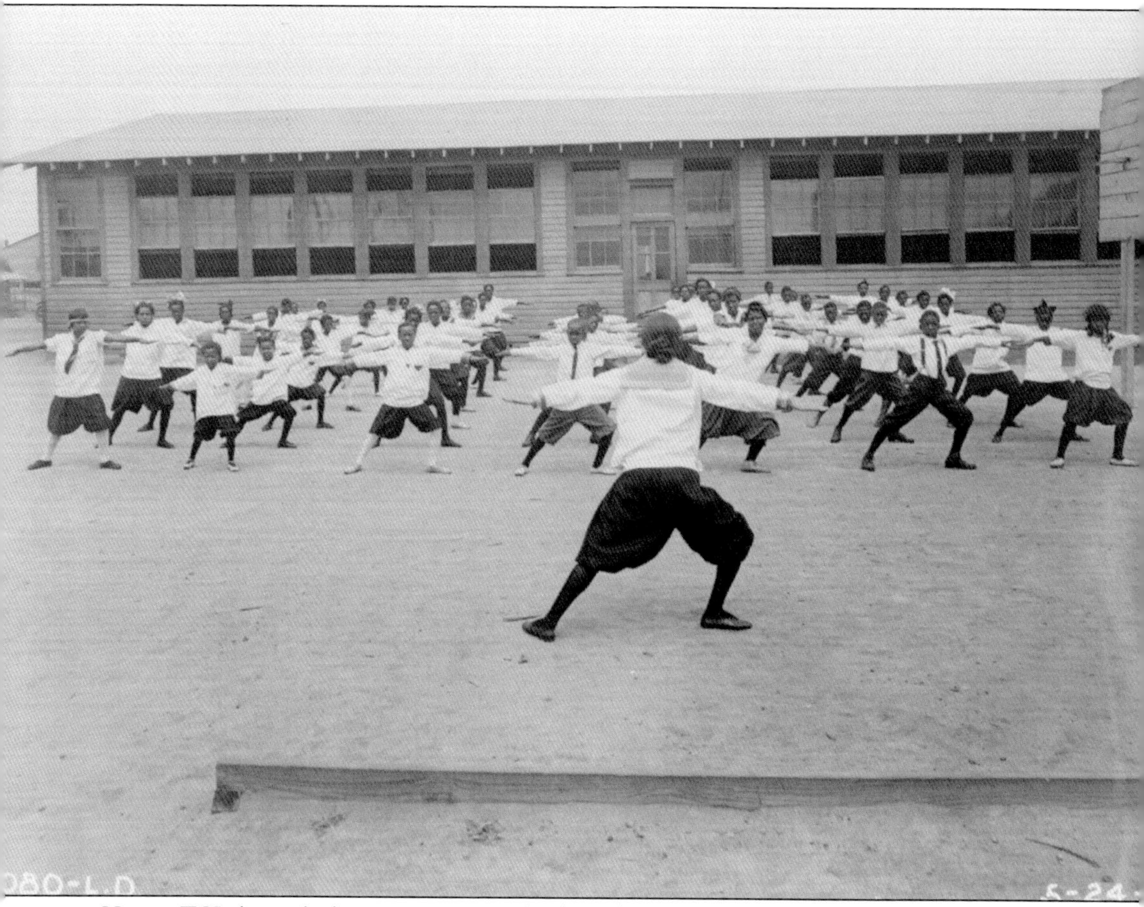

Here, a TCI physical education instructor leads students in a demonstration of physical exercise at the Ensley School for black children. Physical and mental fitness were stressed by TCI's company schools. It was only one part of the company's welfare capitalism model to produce happy working families. (LOC.)

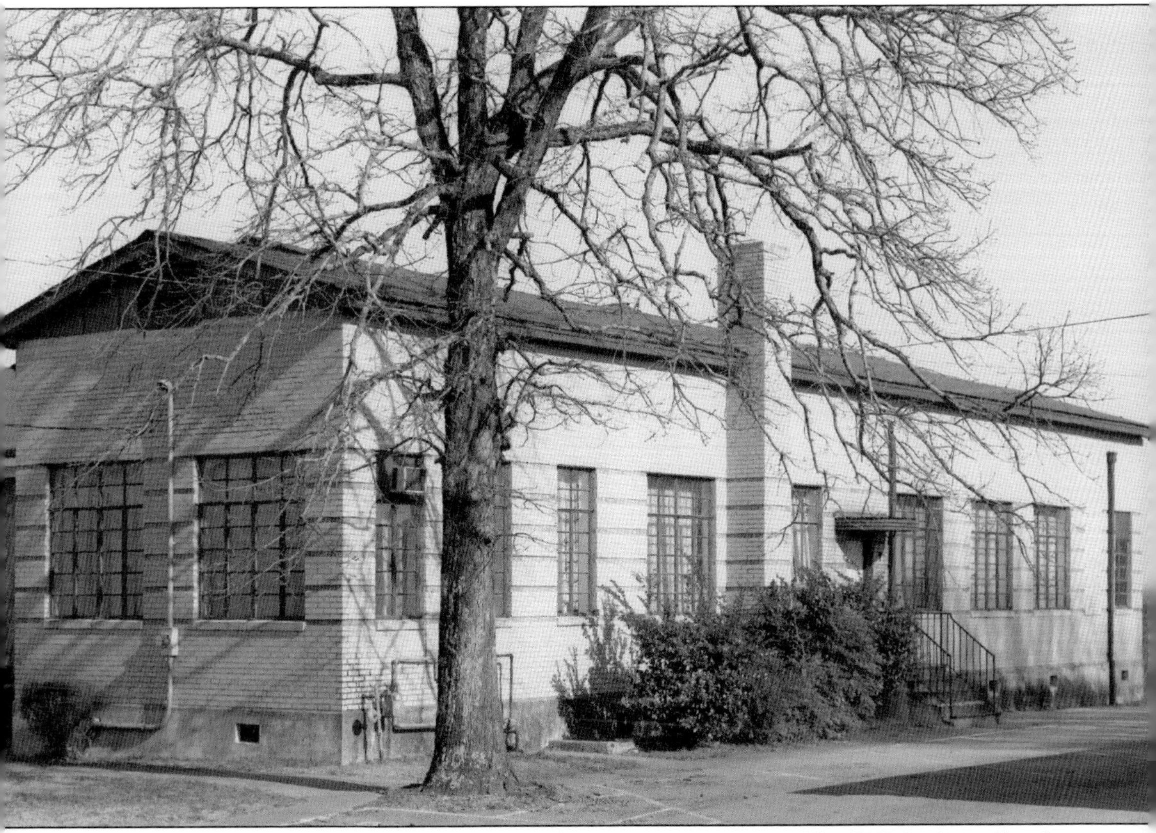

This type of dispensary was common among coal mining communities. This TCI facility offered basic medical care and apothecary services. TCI employees could receive medical benefits, including those offered at the company hospital, Lloyd Nolan Hospital, located in Fairfield. The only requirement was a small monthly fee to join the program. (LOC, HABS.)

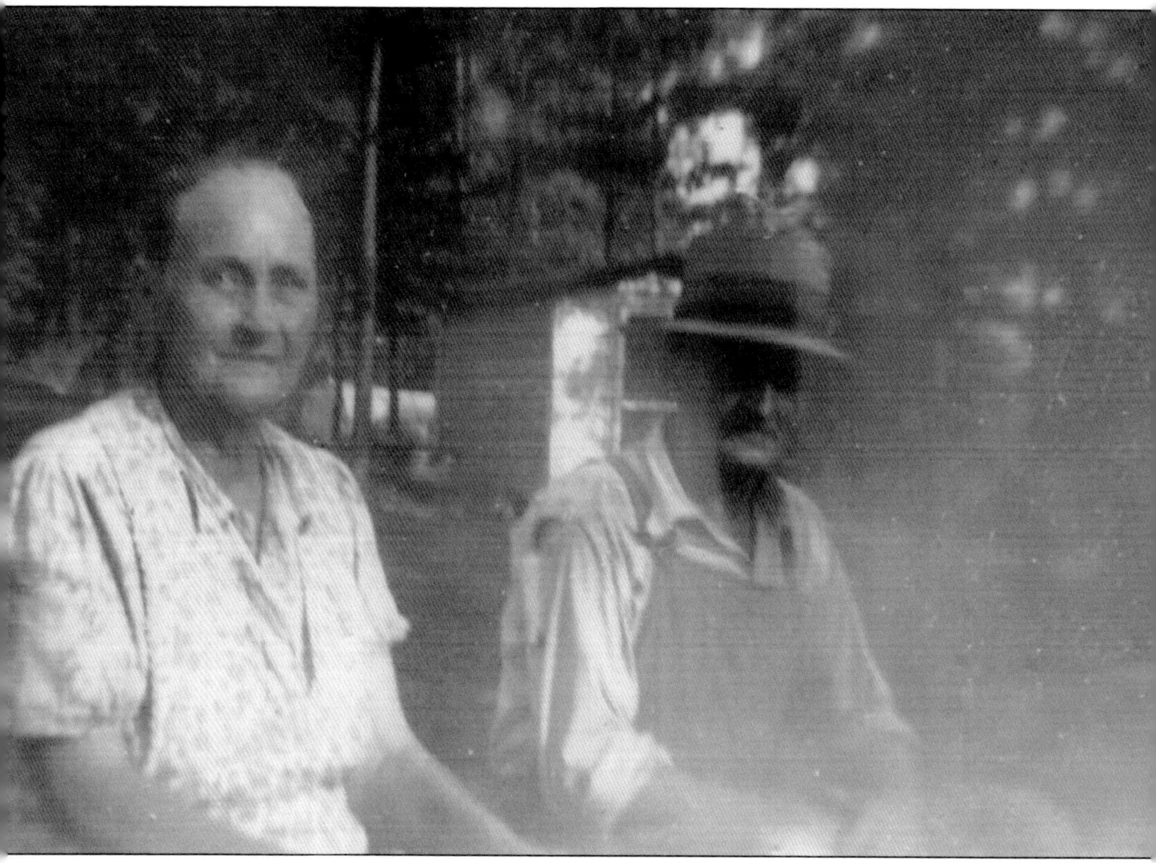

This photograph of an old miner and his wife, Henry Harrison and Nola Brake Goode, from near Mineral Springs, was taken at the end of the heyday of coal mining in Jefferson County around 1947. They had lived to see Brookside, Cardiff, Mineral Springs, Republic, and Watson grow into coal towns right before their eyes. This couple is the author's great-grandparents and part of her inspiration for studying coal miners. (Author's collection.)

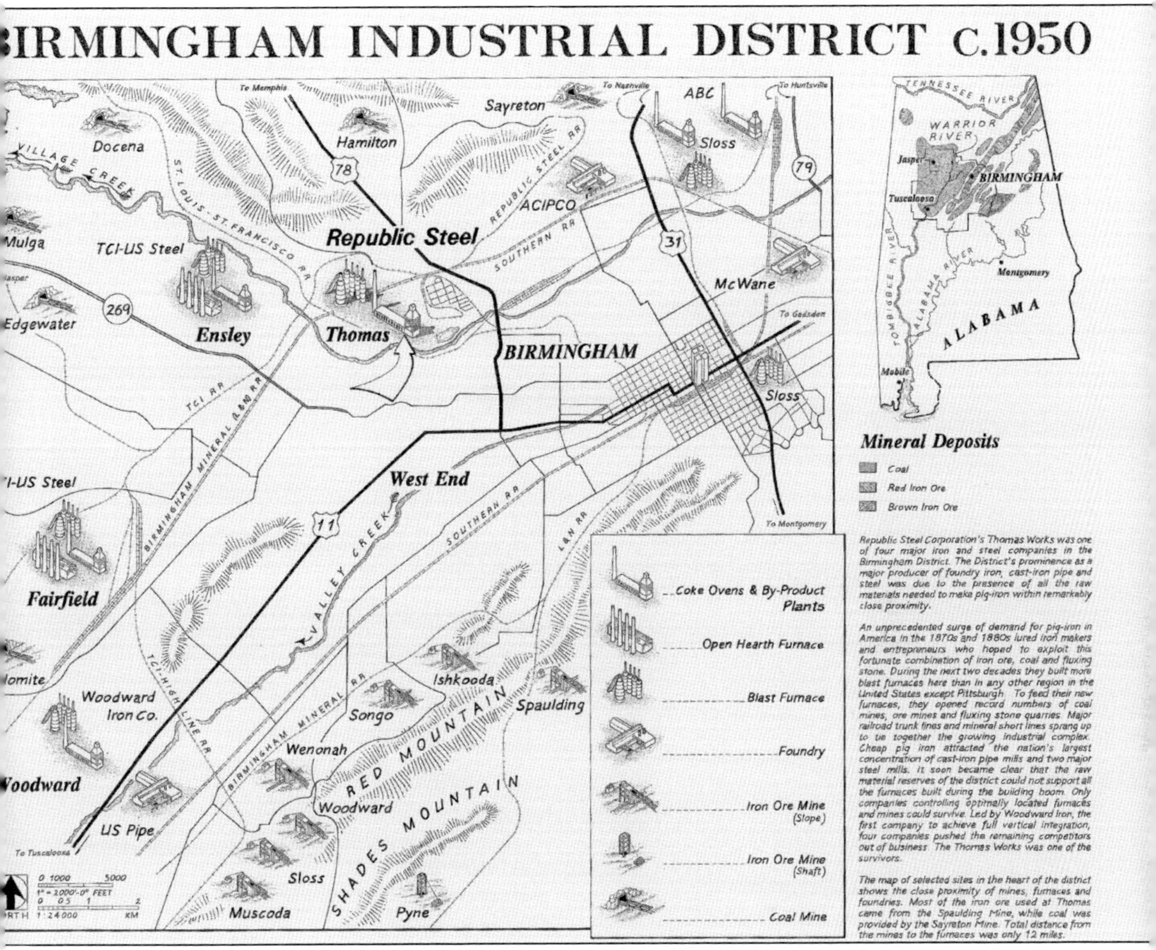

This map shows a dramatically different coal industry than the one that existed in 1900. Only the very largest mines were still in operation, and the miners who were displaced from those closed mines had, by 1950, blended into other facets of Birmingham's economy. (LOC, HABS.)

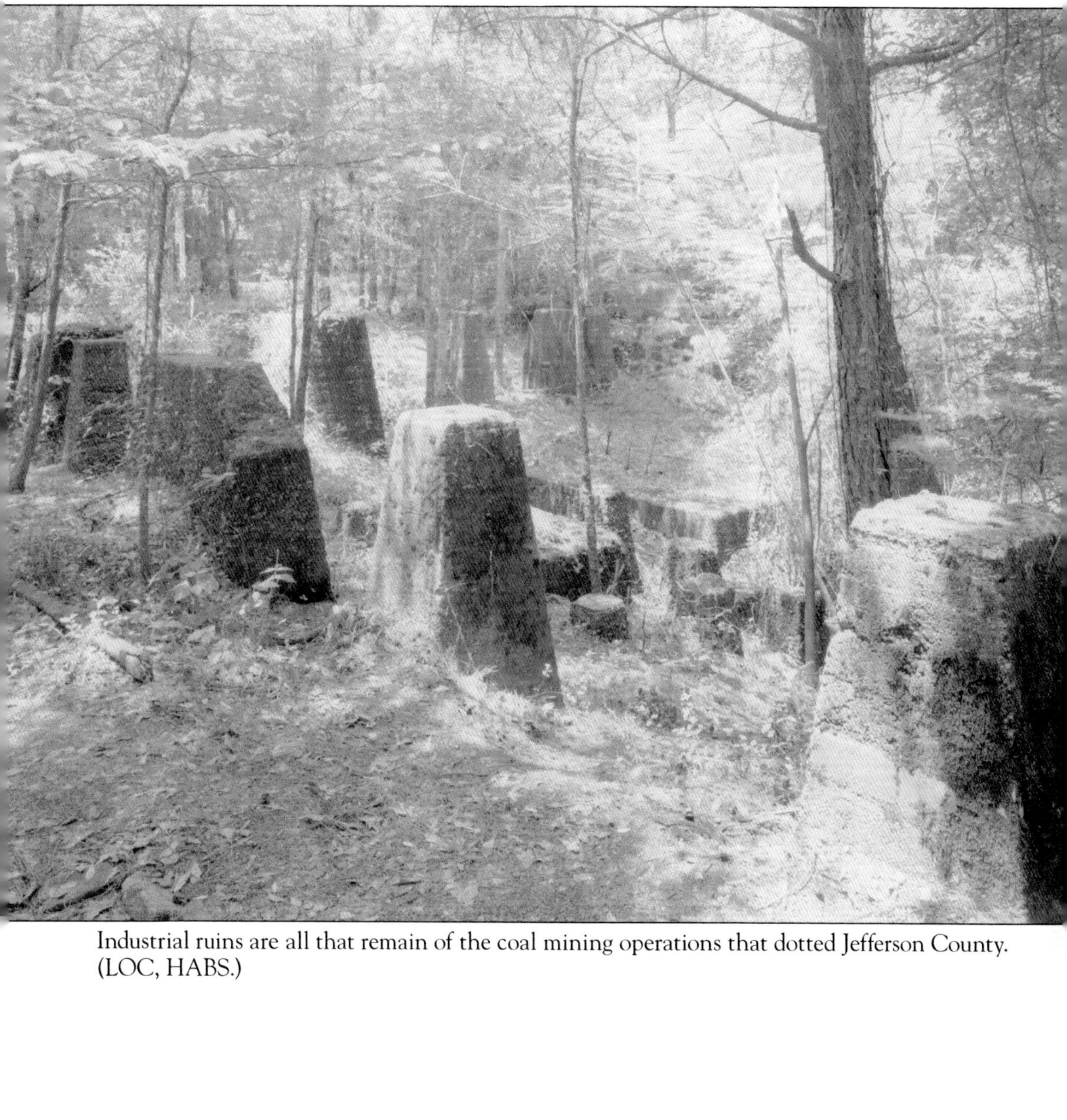

Industrial ruins are all that remain of the coal mining operations that dotted Jefferson County. (LOC, HABS.)

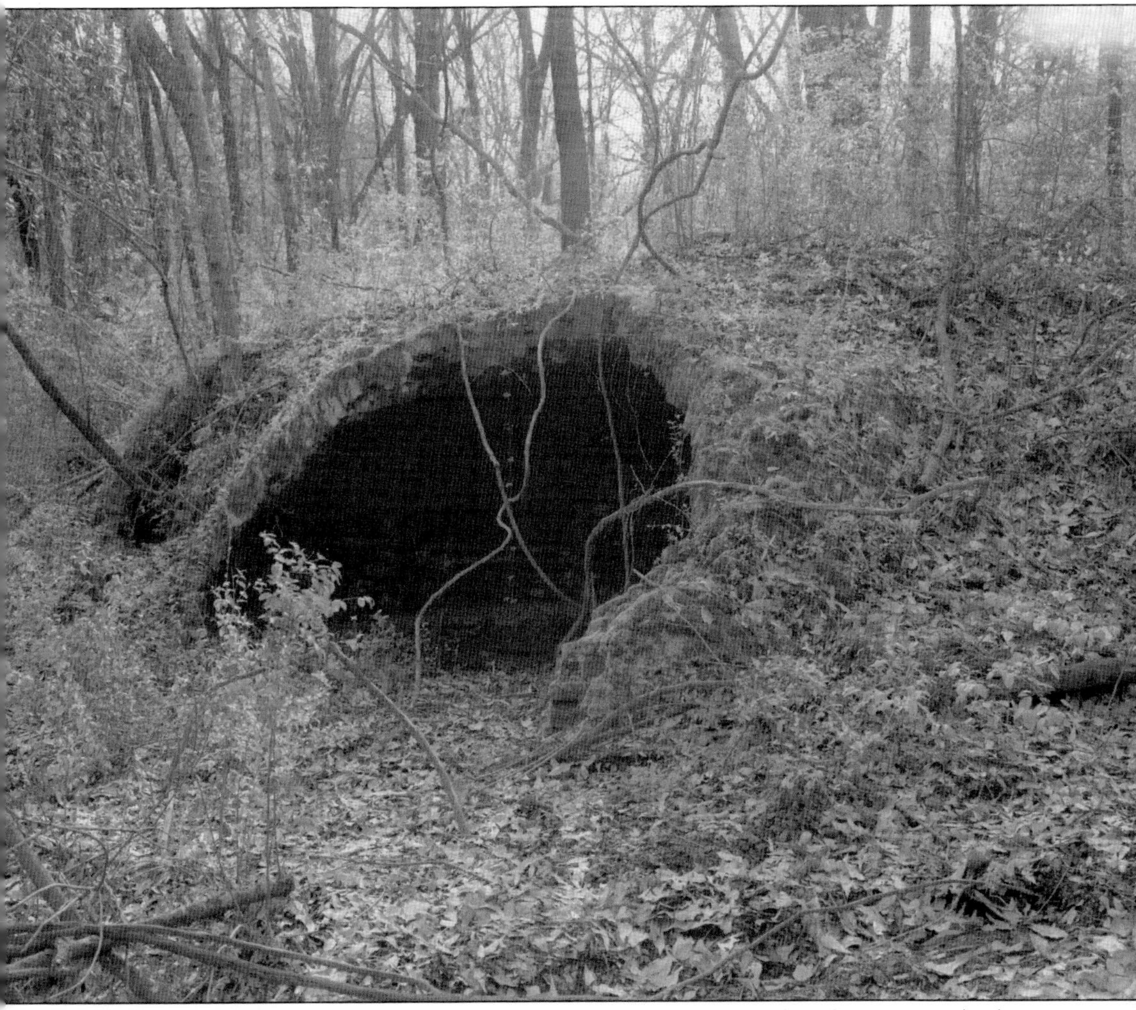

Many coal towns like Fultondale and Brookside are reclaiming their old coke ovens with plans to turn them into parks and pay homage to the coal miners that built the industry of the area. (LOC, HABS.)

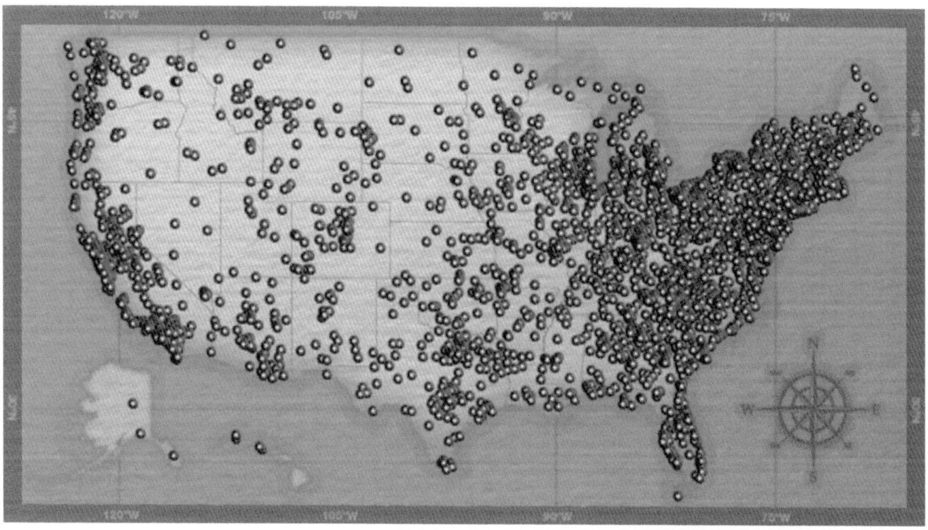

Consistent with our mission to preserve history on a local level, this book was printed in South Carolina on American-made paper and manufactured entirely in the United States. Products carrying the accredited Forest Stewardship Council (FSC) label are printed on 100 percent FSC-certified paper.

MADE IN THE USA